T0269217

Leveraging information technology for optimal aircraft maintenance, repair and overhaul (MRO)

Related titles:

Dynamics of tethered satellite systems
(ISBN 978–0–85709–156–7)
Aimed at engineering students and professionals working in the field of mechanics of space flight, this book examines space tether systems – one of the most forward thinking directions of modern astronautics. The main advantages of this technology are the simplicity, profitability and ecological compatibility: space tethers allow the execution of various manoeuvers in orbit without costs of jet fuel due to the use of gravitational and electromagnetic fields of the Earth. This book will acquaint the reader with the modern state of the space tether's dynamics, with specific attention to the research projects of the nearest decades. The book presents the most effective mathematical models and the methods used for the analysis and prediction of space tether systems' motion; attention is also given on the influence of the tether on spacecraft's motion, to emergencies and chaotic modes.

Introduction to aerospace materials
(ISBN 978–1–85573–946–8)
The structural materials used in airframe and propulsion systems influence the cost, performance and safety of aircraft, and an understanding of the wide range of materials used and the issues surrounding them is essential for the student of aerospace engineering. *Introduction to aerospace materials* reviews the main structural and engine materials used in aircraft, helicopters and spacecraft in terms of their production, properties, performance and applications. The first three chapters of the book introduce the reader to the range of aerospace materials, focusing on recent developments and requirements. The book then discusses the properties and production of metals for aerospace structures, including chapters covering strengthening of metal alloys, mechanical testing, and casting, processing and machining of aerospace metals. The next ten chapters look in depth at individual metals including aluminium, titanium, magnesium, steel and superalloys, as well as the properties and processing of polymers, composites and wood. Chapters on performance issues such as fracture, fatigue and corrosion precede a chapter focusing on inspection and structural health monitoring of aerospace materials. Disposal/recycling and materials selection are covered in the final two chapters.

MEMS for automotive and aerospace applications
(ISBN 978–0–85709–118–5)
Micro-Electro-Mechanical-Systems (MEMS) are miniature devices or machines which integrate elements such as actuators, sensors and a processor to form microsystems. The automotive sector is currently the biggest consumer of MEMS and this market is expected to grow, driven by safety legislation. Emerging applications in the aerospace field will face unique challenges related to harsh environmental conditions and reliability requirements. Part one covers MEMS in automotive applications, including safety systems, stability control and engine management. Part two describes MEMS in aircraft such as navigation systems, devices for health monitoring and drag reduction. MEMS thrusters for nano and pico satellites are also covered.

Details of these and other Woodhead Publishing books can be obtained by:

- visiting our web site at *www.woodheadpublishing.com*
- contacting Customer Services (e-mail: sales@woodheadpublishing.com; fax: +44(0) 1223 832819; tel: +44(0) 1223 499140; address: Woodhead Publishing Limited, 80 High Street, Sawston, Cambridge CB22 3HJ, UK)

If you would like to receive information on forthcoming titles, please send your address details to Customer Services, at the address above. Please confirm which subject areas you are interested in.

Leveraging information technology for optimal aircraft maintenance, repair and overhaul (MRO)

ANANT SAHAY

WOODHEAD
PUBLISHING

Oxford   Cambridge   Philadelphia   New Delhi

Published by Woodhead Publishing Limited, 80 High Street, Sawston,
Cambridge CB22 3HJ, UK
www.woodheadpublishing.com
www.woodheadpublishingonline.com

Woodhead Publishing, 1518 Walnut Street, Suite 1100, Philadelphia, PA 19102–3406, USA

Woodhead Publishing India Private Limited, G-2, Vardaan House, 7/28 Ansari Road,
Daryaganj, New Delhi – 110002, India
www.woodheadpublishingindia.com
www.woodheadpublishing.com

First published 2012, Woodhead Publishing Limited
© A. Sahay, 2012
The authors have asserted their moral rights.

British Library Cataloguing in Publication Data
A catalogue record for this book is available from the British Library.

Library of Congress Control Number 2012943084

Woodhead Publishing ISBN 978-0-08-101642-8 (print)
                     ISBN 978-0-85709-143-7 (online)

Typeset by RefineCatch Limited, Bungay, Suffolk
Printed in the UK and USA

Contents

List of figures and table   ix

List of abbreviations   xi

Acknowledgements   xvii

Prologue   xxi

About the author   xxv

Foreword   xxxi

Introduction   xxxvii

1   An overview of aircraft maintenance   1

    1.1   Aircraft maintenance process   3

    1.2   Maintenance objectives   7

    1.3   Aircraft maintenance strategies   9

    1.4   The maintenance organisations   11

    1.5   Regulatory compliance   12

    1.6   Conclusion   14

    1.7   Notes   14

2   The business of maintaining aircraft   15

    2.1   The aviation MRO market overview   16

    2.2   Customers – who are they?   17

    2.3   Demand and capacity planning   19

    2.4   Service offerings   20

    2.5   Customer orders and contracts   23

    2.6   Order commitment and fulfilment   24

    2.7   Invoicing   25

2.8   Warranty in the world of aircraft maintenance   26

2.9   Customer care   30

2.10 Conclusion   31

2.11 Notes   31

3   Aircraft maintenance paradigm   33

3.1   The paradigm   33

3.2   Life cycle of a commercial aircraft   36

3.3   Airframe maintenance life cycle   53

3.4   Aircraft engine maintenance life cycle   64

3.5   Aircraft components maintenance life cycle   71

3.6   Ground support equipment/fleet (GSE/F)
      maintenance life cycle   75

3.7   Manage materials and logistics   78

3.8   Manage finance   86

3.9   Manage human resources   90

3.10 Manage facilities   93

3.11 Manage continuous improvements   95

3.12 CASS (Continuing Analysis and
      Surveillance System)   97

3.13 Manage environment   99

3.14 Manage information technology   100

3.15 Manage external relationships   104

3.16 Manufacture aircraft parts   105

3.17 Organisation structure   107

3.18 Summary   111

3.19 Notes   113

4   Aviation MRO organisations' challenge to the IT industry   115

4.1   Too many aviation MRO standards and
      lack of them   116

4.2	The conundrum of ownership	123
4.3	Forecasting: does the industry have a crystal ball?	124
4.4	On-wing vs off-wing: the life value cascades	128
4.5	MRP-3 vs MSG-3: the ERP paradigm does not work	130
4.6	Every maintenance activity has to have a Task Card	132
4.7	Many systems to integrate	133
4.8	Summary	137
4.9	Note	137
5	**The IT industry responds**	**139**
5.1	The aviation MRO business and information technology	140
5.2	The era of bespoke systems	141
5.3	The vacuum and the minnows	160
5.4	The big boys get interested	167
5.5	The active vendors	170
5.6	Summary	174
5.7	Note	175
6	**The current aviation MRO IT landscape**	**177**
6.1	The legacy solutions	178
6.2	Best-of-breed solutions	181
6.3	Integrated ERP solutions	184
6.4	The technologies	186
6.5	So many solutions but no holy grail	188
6.6	Summary	190
6.7	Notes	191

7   Leveraging IT and shaping the future   **193**

    7.1   Airworthiness and information technology   194

    7.2   The business view   198

    7.3   The ideal solution   199

    7.4   Is this feasible?   207

    7.5   The nirvana   210

    7.6   Conclusion   212

    7.7   Notes   213

8   Conclusion   **215**

Appendix   **221**

Bibliography   **229**

Index   **231**

List of figures and table

1.1 Aircraft maintenance process   3

3.1 Aircraft life cycle   39

3.2 Airframe maintenance life cycle: aircraft visit   54

3.3 Engine maintenance life cycle: engine
     overhaul/repair   64

3.4 Component maintenance life cycle: component
     overhaul/repair   71

3.5 GSE maintenance life cycle: Ground Support
     Equipment overhaul/repair   75

3.6 IT architecture   103

4.1 MRO system integration   135

4.2 MRO functional integration   136

5.1 A typical CICS login screen   147

5.2 IBM 3270 terminal   149

5.3 IBM S/360   150

5.4 IBM S/370   151

7.1 MRO enterprise: delivering shareholder value   197

7.2 IBM architecture   209

7.3 Some activities of an ideal infrastructure   211

7.4 Expectations of an ideal MRO system   213

Table 2.1 Airline segments   18

List of abbreviations

AC	Advisory Circular
ACARS	Aircraft Communications Addressing and Reporting System
AD	Airworthiness Directive
ADABAS	Adaptable Database System
ADAT	Abu Dhabi Aerospace Technologies
AMC	Air Mobility Command
AMM	Aircraft Maintenance Manual
AMTOSS	Aircraft Maintenance and Task Oriented Support System
AOG	Aircraft on Ground
API	Application Programming Interface
APQC	American Productivity & Quality Centre
AR	Accounts Receivable
ARPANET	Advanced Research Projects Agency Network
ATA	Air Transport Association
BFE	Buyer Furnished Equipment
BoB	Best of Breed
BPRE	Business Process Re-Engineering
CAA	Civil Aeronautics Administration
CAR	Canadian Aviation Regulations
CASS	Continuing Analysis and Surveillance System
CDL	Configuration Drawing List
CFR	Code of Federal Regulations
CICS	Customer Information Control System
CMM	Component Maintenance Manual

CMMI	Capability Maturity Model Integrated
COBOL	Common Business-Oriented Language
CODASYL	Conference on Data Systems Languages
COTS	Commercial Off-The-Shelf
CPU	Central Processing Unit
CRM	Customer Relationship Management
DBMS	Data Base Management System
DFDR	Digital Flight Data Recorder
DoD	Department of Defence
DOM	Directorate of Maintenance
DOS	Disk Operating System
DOT	Department Of Transportation
EASA	European Aviation Safety Agency
EDI	Electronic Data Interchange
EMPACS	Engineering and Maintenance Planning and Control System
EPA	Environmental Protection Agency
ERP	Enterprise Resource Planning
FAA	Federal Aviation Authority
FAR	Federal Aviation Regulations
FIM	Fault Identification Manual
FORTRAN	Formula Translation Language
FRM	Fault Reporting Manual
GAMCO	Gulf Aircraft Maintenance Company
GSE	Ground Support Equipment
GSF	Ground Support Facilities
HAECO	Hong Kong Aircraft Engineering Company Limited
HCL	Hindustan Computer Limited
HP	Hewlett Packard
HR	Human Resources
HTML	Hyper Text Mark-up Language
IATA	International Air Transport Association
IBM	International Business Machines

III-TRM	Integrated Information Infrastructure – Technical Reference Model
ILC	International Logistics Centre
IMS	Information Management System
IP	Intellectual Property/Internet Protocol
IPC	Illustrated Parts Catalogue
JAL	Japan Airlines Ltd
JAR	Joint Aviation Regulations
JCL	Job Control Language
JEMTOSS	Joint Engine Maintenance and Task Oriented Support System
KL	Kuala Lumpur
KPI	Key Performance Indicator
LCC	Low Cost Carrier
LRU	Line Removable Unit
MAE	Malaysian Airline Engineering
MAS	Malaysian Airline System
MCC	Maintenance Control Centre
MEL	Minimum Equipment List
MEMIS	Maintenance and Engineering Management Information System
MODs	Modifications
MPD	Maintenance Planning Document/Database
MPRB	Maintenance Planning Review Board
MPWG	Maintenance Planning Working Group
MRB	Maintenance Review Board
MRP	Manufacturing Resource Planning
MS	Microsoft
MSDS	Material Specification Data Sheet
MSG	Manufacturer and Supplier Group
MSI	Maintenance Significant Item
MVS	Multiple Virtual Storage
NDT	Non Destructive Test
NFPA	National Fire Prevention Association

NPL	New Programming Language
OEM	Original Equipment Manufacturer
OHSA	Occupational Health and Safety Act
OS	Operating System
PANAM	Pan American Airlines
PC	Personal Computer
PIREP	Pilot's Report
PMA	Parts Manufacturer Approval
PwC	PricewaterhouseCoopers
QANTAS	Queensland and Northern Territory Aerial Services
RCM	Reliability Centred Maintenance
RFID	Radio Frequency Identification
SABRE	Semi-Automated Business Research Environment
SAE	Society for Aerospace Engineers
SB	Service Bulletin
SCEPTRE	System Computerized for Economical Performance, Tracking, Recording and Evaluation
SFE	Supplier Furnished Equipment
SGML	Standard Generalized Markup Language
SIA	Singapore Airline
SISP	Strategic Information Systems Planning
SNA	Systems Network Architecture
SOA	Service Oriented Architecture
SRM	Supplier Relationship Management
TCDS	Type Certificate Data Sheets
TCP/IP	Transfer Control Protocol over Internet Protocol
TELEX	Telegraph/Teleprinter Exchange
TOGAF	The Open Group Architecture Framework
TSN	Time Since New
TSO	Time Since Overhaul

UAL	United Airline
VM	Virtual Machine
VMS	Open Virtual Memory System
VSAM	Virtual Storage Access Method
VSE	Virtual Storage Extended
VTAM	Virtual Telecommunications Access Method
WBS	Work Breakdown Structure
XML	Extensible Mark-up Language

Acknowledgements

The Divinity within me perceives and adores the Divinity within you.[1]

I've heard it said that writing a book is like giving birth, except the pain lasts longer.

While I have no way of establishing the truth of that saying, all I know is that it took much longer than nine months to develop the ideas contained here and more than two years to write it all down. As to the pain, I think I was quite good at sharing that – you had to be there. And many of my closest friends and family were.

I am humbled by the support, encouragement, and sheer pulling up of bootstraps that many of my friends and family engaged in to help me along the way, and I must acknowledge that there are many without whom all these ideas would have remained in my head, and I would have forever been that boring dinner companion who discussed the finer points of information technology and aircraft maintenance over a nice glass of Cabernet Sauvignon.

Before I go any further, I must acknowledge the contribution and support of my wife, Naintara; and children, Nishant, Hersh and Chetan. Whether it was because they had to listen politely to the above mentioned dinner conversations, or because they believe that there are many out there who would be fascinated by the subject of complex MRO, I have never quite established; but it was they who made sure this book got written. They reminded me constantly of publishing deadlines, undertook editing and proof reading of many chapters, and allowed me to retreat from the world to write my book, especially when in the midst of family gatherings, holidays and social events. For this I am eternally grateful.

Of the many friends and colleagues who set me on the path to make this book a reality, I must first acknowledge my friend and fellow author, George Mathew. Being totally ignorant of the publishing industry and the torturous process of finding a suitable publisher, it was George who provided practical and material guidance throughout and provided thoughtful tips and techniques to speed up and streamline the more tedious tasks of creating a glossary and index. Without the help and support so generously provided by him, I could have spent weeks on the index alone.

I need to thank also my friend of many years, Sudhakar Murthy, CEO of Acore Group, who not only wrote the initial review along with Michel Parsons of Oracle and Kernail Singh of Jet Airways, but also took time off his busy schedule to write the Foreword. I would also like to thank and acknowledge the important role of Kaldip Singh of Malaysian Airlines who asked me to write a book.

I am extremely grateful to Thomas Deluca of RAMCO, who provided the history of MAXI-MERLIN based on his days of working with USAir. His only demand was that he gets one of the first signed copies of this book.

Catherine Davies, the Editor of *AviTrader MRO & Weekly Headlines*, provided a morale boost by publishing my interview with her. I thank her for including me in the article, which had many prominent aviation MRO experts interviewed.

I would also like to thank my colleagues, who indirectly contributed towards this book: Mark Pemberton, Rob Powell, Nishant Balakrishnan, along with many others at IBM.

Finally, I would like to acknowledge the publishing team, whose perseverance – in spite of many slipped deadlines –

was essential for the completion of the book. I thank Dr Glyn Jones, Jonathan Davis and Vicki Hart of Chandos Publishing for managing the progress of the book; Annette Wiseman of RefineCatch Ltd for her diligent editing; and George Knott of Chandos Publishing for providing the cover of the book.

There are many whose names I have not mentioned here, but who have in one way or another helped me along the way. I am grateful for their support and encouragement.

Last but not the least I thank the organisations, which gave me the opportunity to learn, practice and execute my aviation MRO skills: Gulf Air, Bahrain; ADAT, Abu Dhabi; Emirates Airline, Dubai; and IBM.

Note

1. Finnegan, Dave. *The Zen of Juggling.* Jugglebug, 1993.

Prologue

There was neither non-existence nor existence then;
there was neither the realm of space nor the sky which
is beyond. What stirred? Where? In whose protection?
.... Whence this creation has arisen – perhaps it
formed itself, or perhaps it did not – the one who looks
down on it, in the highest heaven, only he knows – or
perhaps he knows not.[1]

One does not know what one is capable of until somebody
else points out to him what he can do; especially what
boundaries he can break. Consider the Indian mythical
Hanuman, the monkey god, who did not know he could fly
until he was explicitly told that he could.

I was in Kuala Lumpur, Malaysia, in a consultancy
engagement with Malaysian Airlines. I had finished my
assignment of creating a blueprint on MRO (maintenance,
repair and overhaul) for their Engineering Department. This
department was on the verge of converting to a separate
business entity and was looking for a strategic roadmap for
Information System implementation.

I presented the outcome of my study to the middle and
senior managers of Malaysian Airline's Engineering
Department. Most of them liked the presentation and some
of them did not as it did contain some mild criticism of their
current way of working and approach to information
technology. However, the audience of about forty listened
carefully as I went through almost fifty slides. At times the

intensity of concentration was very high and sometimes I had to wake some people up by asking specific questions. However, the most important aspect, from my perspective was that the senior management sat through the entire two hour presentation and there were lots of lively and interactive questions.

It was later, though, that Kaldip Singh (the IT Manager responsible for Malaysian Airline Engineering) and I, along with other friends and colleagues, met up in a pub in downtown KL and had a candid discussion about the presentation. He said that during the presentation a thought crossed his mind that I must have a published a book on the subject. Later, according to him, he did search for the 'non-existent' book without success. Over a couple of beers, Kaldeep suggested that I should write a book on the subject.

I had never considered myself as an author but the thought stayed with me, which was later on buttressed by my wife Naintara and many of my colleagues. And then I met George Mathias, a colleague of my wife in Infosys. He had been writing a book for a while. Over the dinner table, when we started talking about his book, he at some point turned the course of the conversation to me writing a book. I was not sure. I dodged the question by revealing my well-known ignorance of the world of authors and publishers. I am an avid reader but never thought of myself as the one who could actually write a book. In my view it required a special talent for a way with words, and being technically orientated, I did not know how to express myself beyond talking about specifics and being concise; I do write a lot of reports and proposals, etc.

As the conversation progressed, George more or less convinced me that I did have a book in me and he would be happy to introduce me to his publishers, which he did the very next day. I was apprehensive, at the start. The publisher,

Dr Glyn Jones of Chandos, asked me to submit a proposal. Initially I thought the task would be daunting but my skills in writing proposals came in handy. I submitted the proposal and it was accepted after being reviewed by three of my colleagues, from different countries and companies. I was then sent a contract. I was so enthused that my instinctive reaction was to sign the contract and send it back. But my other instinct of rigour prevented me from doing so. Naintara and I along with an IBM lawyer in the background went through the contract with a fine toothcomb and sent back to Chandos for review. After some negotiations, we agreed on the terms of the contract and I had signed up myself to write the book.

After signing the contract, it dawned on me that I did have to write. I went through the tomes of documents that I had collected over the years and went into deep introspection. But it was hard translating my thoughts into words. Somehow, the thoughts were so colourful but the words were not. I did not want to write a dry technical manual (I have done many of those) but something that spoke from not just the mind but also the heart. The writer's block was there and I was unable to move beyond the Table of Contents until I decided to take some days off and went to Bali. It must have been either the persuasion from Naintara, or the lovely atmosphere around the beach overlooking the Indian Ocean, or having not to think of my daily job, or a combination of all these that did it. I started writing this book, chapter by chapter. The words came, which represented my thoughts and I kept typing for the next few months.

Note

1. Rig Veda (10.129) Nasadiya. Translation by Wendy Doniger O'Flaherty for Penguin Classics, 2005.

About the author

Anant Sahay has over 30 years of professional experience in information technology for the aviation sector. Over this period, he has been associated with a number of airlines and aircraft maintenance repairers (Original Equipment Manufacturers or OEMs), either as a technical resource or as an industry consultant. These include: Emirates Airline, Gulf Air, ADAT (Abu Dhabi Aviation Technology), Malaysian Airlines, Air China, Rolls Royce, Bombardier, Embraer, Saudi Arabian Airline, Air New Zealand, Qantas, and Eva Air among others.

With a deep experience and insight of the aviation MRO industry, complex processes and the regulatory compliance required by the industry, his career has seen him travel over a million miles to more than 45 countries, working in strategic planning, execution and delivery of IT solutions for aircraft maintenance and engineering, including data warehousing and systems integration. These solutions encompass bespoke as well as packaged solutions from leading vendors

Anant has a Production Engineering degree and later received a PGDBM (Post Graduate Diploma in Business Management) from XLRI, in Jamshedpur, India. He is a PMI certified Professional Project Manager as well a TOGAF Certified Enterprise Architect. As a certified IBM industry consultant (with skills in management, consulting and delivery), Anant has engaged with C level executives and attended senior executive and steering committees. He has actively set up partnerships and alliances with large software vendors and other business partners. He has established competency practices using best practices and processes based on CMMI, TOGAF and other relevant standards and specifications.

Anant's first major MRO project involved design and development of a bespoke technical defects analysis and control application for Gulf Air Engineering. Gulf Air was operating L1011 and B737 aircraft at the time. The defect management system was manual. Since the organisation and consequently the fleet were expanding, and the routes were extending, it became imperative for Gulf Air Engineering to automate the defect management and reporting process. Anant designed, developed and implemented the system with a team of three programmers and analysts, which picked up the defects from PIREP logbooks and used various categories to report technical defects to the regulatory authorities.

This was followed by the design and development of bespoke DFDR and black box analysis application for Gulf Air Engineering. Gulf Air had outsourced the DFDR analysis work to Rockwell in the UK. However, it was beginning to get very expensive. Therefore, Gulf Air decided to develop a bespoke application based on a software package available on the VAX/VMS platform. This application was to read the black box recordings and provide performance of the systems graphically. Anant designed, developed and implemented the system for Gulf Air Engineering and Flight Operations.

Anant also designed and developed a bespoke Technical Records System for Rotables and Repairables for Gulf Air Engineering. HAECO of Hong Kong used this to maintain Gulf Air's L1011 aircraft. Later, Gulf Air decided to move the maintenance work to GAMCO. Since HAECO managed the technical records on its proprietary system, an interim Technical Records Management system was required for transition. Anant designed, developed, and implemented the system in less than three months with a team of five programmers and analysts. Though the application was developed as a temporary solution it was actually used for seven years until all the L1011s were decommissioned.

During the late 1980s, Gulf Air was expanding its fleet by introducing B767 and A320 aircraft and decided to implement a package solution to support the maintenance of these aircraft. After several evaluation cycles they selected SCEPTRE from Republic Airline (now merged with North West). SCEPTRE boasts superior configuration management functionality. However, it uses an IMS database, which was not a strategic platform for Gulf Air. After reviews and discussions, Gulf Air decided to modify SCEPTRE to run on ADABAS. PANAM Consultants from Florida, USA were engaged to carry out the modification.

Anant, with a team of seven programmers and analysts, fine-tuned the software and implemented the application in phases. The first phase was implemented for Gulf Air and the other phases were implemented at ADAT, Abu Dhabi, once their maintenance facility was ready for use.

Later ADAT, formerly GAMCO, decided to move to a newer technology and decided to implement MAXI-MERLIN from USAIR (currently the software is owned by SABRE). Anant was seconded to GAMCO as a System Implementation Manager to manage the implementation of MAXI-MERLIN over two years, with a team of thirty software engineers.

In the early 1990s, Anant joined Emirates Airline. Emirates' core MRO application was EMPACS, which was a function rich mainframe-based system. However, it did not have a SPEC2000 interface for B2B spares procurement, Shelf-Life control functionality or an interface to ACARS for monitoring defects while the aircraft was in flight. Anant designed, developed and implemented the subsystems to fill these gaps with a large team of programmers and analysts.

During this time, Emirates Engineering was in the market looking for an Aviation Production Planning and Control System for Heavy Maintenance. The extant application had minimal functionalities and needed significant enhancement. Since no suitable COTS package was found, it was decided to develop a bespoke application. Anant with a team of five software engineers, designed, developed and implemented the PP&C application, which was linked to Boeing's documentation system and was compliant with ATA 2100 standards for Task Card management and printing.

Emirates Engineering used EMPACS (Engineering Maintenance Planning and Control System) developed by Cathay Pacific. Emirates had bought the source code, and under Anant's leadership a local team was deployed to maintain this. In spite of significant development work in-house, a review of the CX system revealed that EK's EMPACS lacked a certain critical functionality, which CX had already deployed. Anant evaluated CX EMPACS and, based on the gap analysis, prepared a business case for an EMPACS upgrade, which was approved by EK Engineering.

Anant negotiated the deal with CX and managed the upgraded program.

Anant pioneered a Business Intelligence and Information Management solution to meet the analytical reporting requirements of EK Engineering. The solution provided a platform to integrate data from legacy systems and analytic reporting, and replaced around 750 standard reports put out by EMPACS at various times. This system grew from eight to eight hundred users within two years of its implementation.

Emirates Airline launched a program for Strategic Information System Planning for the entire organisation and Anant was responsible for developing the Information Systems strategy for EK Engineering. He delivered a five-year strategy and roadmap for MRO systems with a set of initiatives which were subsequently costed, budgeted and implemented.

In parallel with the SISP (Strategic Information Systems Planning) a BPRE (Business Process Re-Engineering) program for EK Engineering was initiated. Anant led the BPRE team.

Anant was part of the Boeing 777 acquisition team for Emirates Airline and spent one month in Seattle at Boeing's Everett facility. His main task was to identify all the software and hardware to support induction of new aircraft into the Emirates fleet. Anant supported the initial provisioning for aircrafts as well as in setting up systems for the new spares parts warehouse.

MAE (Malaysian Aerospace Engineering) is a wholly owned subsidiary of MAS (Malaysian Airlines). MAS launched a business transformation initiative, which involved implementation of ERP and MRO applications. IBM was engaged to prepare the blueprint for the overall initiative and Anant was responsible for the MRO part of the Blueprint.

Anant is a naturalised Australian citizen and lives in Sydney, Australia. He can be contacted at *anant@sahay.net.*

Foreword

It gives me great pleasure and honour to write a foreword for a book which deals with a subject that covers the entire spectrum of my professional life as an Aircraft Engineer in Aircraft Maintenance . . . from an airline to an aircraft manufacturer's perspective including the engine and component manufacturer with respect to MRO activity. Even though I am now the CEO of ACORE Group, I still call myself an aircraft engineer. ACORE stands for Aircraft Component Overhaul Repair Engineering . . . currently involved in bridging the gap and creating a seamless transition to a Long Term Total Maintenance Support for Airlines. Well . . . this is the trend for the future and a way to control and reduce maintenance costs in the long run, thereby increasing the revenues and profits!

In the earlier days of my career in Air India when there were no computers, we did manual handling of aircraft maintenance – monitoring, planning, materials, repair processes and controls. It would take a long time for the information to be exchanged between the various departments involved in these activities, resulting in delayed actions and increased costs. The fact that the maintenance and fuel costs make up the majority of the total airline costs, resulting in higher flying costs, made it difficult for people to fly in those days. So what did IT bring to make our life simple and rich along with the vast technological changes that took place in aviation? It introduced complex maintenance systems.

My mentor Mr. Bashir Abdel Hadi, an industry veteran with several years of running an MRO outfit successfully shares his valuable experience with me. I am thankful to Bashir for sharing his analysis, views and thoughts from a total perspective. According to him, the MRO business in general is facing major challenges. MRO revenues are expected to grow at rates lower than the growth of the aviation industry in general because of several factors, the fleet is getting younger, the new generation aircraft require less maintenance at longer intervals, engines and components are much more reliable, and airlines are becoming smarter in managing their maintenance needs. The declining demand will result in higher capacity offered, and consequently higher competition.

He goes on to say that the story on the cost side is no brighter. In addition to the continuous investment in infrastructure in order to keep up with the development of equipment, most MRO business models are built on high overhead costs with more than one third of costs related to labour and the another third related to material, leaving very small room for the controllable costs.

With the reduced demand and the increasing costs, efficiency becomes vital, and improvement in efficiency can only be achieved through advance HR practices and effective IT solutions.

Information technology is a business function, which uses process and technology regulations to define, manage and share master data across the organisation. In addition, it facilitates the major tasks of planning, executing and controlling within an organisation. A well-structured IT system sets objectives and selects the best course of action to meet these objectives through the coordination of human and intangible resources. In addition, it ensures monitoring and measuring progress regularly so as to identify variances from original plans and to take corrective action, where necessary, to meet objectives.

The role of information technology in Airframe and Engine MRO became very critical with technological advancements in the last three decades. Maintaining the operational capabilities of aircraft/engines and their accessories according to the guidelines recommended by the manufacturer, the requirements established by the regulatory agencies and the standards established by a specific branch of the government, is an expensive proposition. So, this burden created by the time consuming task of managing maintenance operations was greatly facilitated by an effective software-driven IT network in managing certain facets of the maintenance enterprise that integrates the entire maintenance function. This cost effective, software-driven IT system seamlessly integrates scheduled and preventative maintenance with parts inventory. The scheduling function shall ease the task of compiling critical inspection paths to minimise turn around time.

However, selecting the most cost effective software requires establishing a plan that will assure that the cure provides a solution and not exacerbate the problem. Functional flexibility in an aircraft maintenance software program allows the customisation of databases permitting various departments to share one productivity package. In order to obtain the best system, some airlines use separate schedulling and maintenance software programs, as long as the two programs can be integrated.

The parts inventory or spare inventory management portion have built-in controls consistent with those utilised within a maintenance branch. Once stock quantities and ordering points are determined, a system can automatically generate recommended orders that can be approved and forwarded electronically to suppliers. Stock control and order quantity is a critical functionality. A system that can be meshed with vendors systems will increase the speed with which parts can be replenished thereby reducing the amount required in inventory. This interaction reduces the overall cost of a maintenance operation.

The recent use of Radio Frequency Identification (RFID) in maintenance systems has become an important function. The use of barcodes, scanners and readers that are integrated with the software increases the effectiveness of inventory management. RFIDs increase efficiency, help track parts usage and are an aid in inventory control.

It is a proven fact that MROs with effective IT management have achieved a wide range of benefits such as accurate reporting, effective monitoring and analysis of activities throughout business processes, a comprehensive view of customers, products and supplies and reduced errors. This leads to very high degree of productivity and thus very cost effective maintenance within the framework of required safety and airworthiness standards.

Bashir Abdel Hadi strongly feels that in view of the complexity of MRO activities and the importance of turn around times, MRO is a great candidate for an ERP system. Unfortunately there are more failures than success stories in ERP implementations in the MRO arena. In many cases, after several years of implementation agony, the outcome is utilising a small portion of the intended system and resulting in increasing the workload rather than reducing it.

So the future is for a plug-and-play system that is able to prove to the industry that it has major positive impact on efficiency.

Having worked with Anant Sahay in Gulf Air in the very early stages of integrating and implementing IT packages with the existing MRO functions and going through the agony of customising it in different departments, made us understand each other very well. Anant's sound engineering background and a clear understanding of the logics and sensitivities of an aircraft MRO industry made him one of the best hands available to successfully develop, integrate and customise IT solutions in an airline environment. It is this knowledge, background and experience which makes him the ideal author of such an interestingly delicate subject where the interface takes away the major creative aspect of the whole game. Anant has championed this important area of the business wherein he can combine his IT knowledge and experience with the airline MROs very effectively.

I am confident that the book will be very interesting and valuable for not only for aircraft engineers like me but also for the young IT engineers getting to know about this industry and working in it in the future. It would be a good textbook for any course and will also serve as a reference book for all those involved in the aircraft MRO business that includes airlines as well as manufacturers.

Sudhakar Murthy
Dubai
20 May 2012

Introduction

Where there is creation there is progress. Where there is no creation there is no progress: know the nature of creation.[1]

Industry overview

In the early days of flight, aircraft were flown the way one drives a car. The pilot was the mechanic, the driver, and mostly the most important passenger. Airworthiness was the sole responsibility of the pilot, but things did not remain that way for long. Once aircraft became a mode of paid transportation and therefore subject to formal scrutiny, and even legislation, the responsibility of maintaining the airworthiness of a passenger airliner shifted to the ticket-issuing organisation: the airline itself.

In 2010, the IATA (International Air Transport Association) declared that 'Flying is safe'. This is very well supported by the fact that there was only one accident for every 1.6 million flights in that year. Overall 786 people were killed in 23 accidents. This is lower than almost all mechanised and motorised means of transport. This is not a trivial achievement.

The aviation industry has worked very hard right from the inception of commercial air travel. It took a while to get there, but it has become a showpiece achievement brought about with unique cooperation between the governments and aircraft manufacturers, especially in the USA.

At the heart of all this was the concern for passenger safety. Psychologically, flying (that is travelling through thin air), is the riskiest proposition for human beings. Just the idea of hurtling through a vast space thousands of metres above land, among the clouds, is extremely daunting. We still have many amongst us who suffer from fear-of-flying. This paranoia, combined with the inability to suppress the desire to go from one place to other in a jiffy, brought the industry and the government together to devise extremely sophisticated, complex and rigorous processes to ensure that an aircraft does not fall out of the sky due to failure of any of its components, systems or structure.

All this has been achieved because of the rigour and discipline put in place for maintaining a commercial aircraft. It is easier said than done. To maintain such a level of safety, a complex and disciplined set of process are used. We shall discuss these in detail in the subsequent chapter. However, I will say that it is all about airworthiness. A single minded effort to achieve airworthiness is what makes air travel safe.

Airworthiness

There are three ways of looking at airworthiness:

1. the common sense aspect
2. the legal aspect
3. the technical aspect.

In this chapter and throughout the book we will concentrate mainly on the last and final one, the technical aspect of airworthiness. However, it is important to understand the other two aspects and to recognise their importance in the industry and to fully appreciate the import of the terms.

The general definition of the term airworthiness is: 'Being in a fit condition to fly.' This can be further elaborated as an aircraft safe and fit to fly. This applies to all the perspectives mentioned above.

A passenger, a cargo/freight operator and a lay person are satisfied with the commonsensical perspective of airworthiness. In the case of a passenger, as long as the aircraft is authorised to fly and is available for him to use, the aircraft is airworthy. Similarly, an outsider who sees the aircraft flying, taking off or landing (whether cargo or passenger aircraft), considers the aircraft airworthy. A passenger does not ask for any proof of airworthiness before boarding an airplane. He simply takes it for granted that since the aircraft is ready to board and he has heard the instruction to do so, the aircraft is safe and fit to fly. So does the cargo or freight operator. For an external viewer the visual confirmation is enough and furthermore, since it does not concern him, he does not need any proof.

However, the legal perspective requires more elaborate measures. It needs evidence. An aircraft is airworthy only if there is evidence that explicitly states that it has been signed off by an authorised representative of the aircraft operator, who is liable for safety and fitness for flight of the aircraft. This is, in civil aviation parlance, an Airworthiness Certificate. In other words, legally an aircraft is deemed safe and fit to fly if it has an Airworthiness Certificate. The process of issuance of the certificate and how information technology does and can support and facilitate the process will be discussed in detail later in this book.

Regulatory authorities around the world have defined airworthiness and it is their published documents which require compliance by the aircraft operators depending on their geographical locations and operations. Some examples of these are:

- The US *Federal Aviation Regulations*, Part 21, §21.183(d) has a procedural definition of *airworthy*.
- Airworthiness is defined in JSP553 Military Airworthiness Regulations (2006) Edition 1 Change 5.
- In Canada *Canadian Aviation Regulations*, CAR 101.01, Subpart 1 – Interpretation Content last revised: 2007/12/30.

It is worthwhile noting that every passenger and cargo operator has the right to inspect these certificates. In the recent past, it used to be the practice to affix a 'type' certificate next to the door of the aircraft but lately this practice has fallen into disuse. However, a passenger or a freight operator was never privy to airworthiness certification before each flight. This was more due to logistics constraints than availability of the certificate.

Currently the aircraft operators are in a position to publish these certificates on the web so that any interested passenger or freight operator can sight them. This will not necessarily improve efficiency in any way; however, the cost of taking this step is so small that it would be a good idea to introduce this practice to add to the awareness of the operators and confidence of passengers.

The technical perspective of airworthiness requires a more detailed discussion. It assumes that an aircraft is safe and fit to fly if all the critical components: aggregates, systems and structures, are functioning as designed. The implication here is that legally the aircraft has been type certified, i.e. all its components are designed and manufactured to safely fly the aircraft.

One of the greatest outcomes of Boeing's 747 program (apart from the wonderful aircraft itself) that changed the travel industry, was the standardisation of the aircraft maintenance process. The objective of this process, enshrined in the document created by MSG (Manufacturer and Supplier Group), was to provide a standardised logic to arrive at the airworthiness of an aircraft. This standard was twice enhanced and continues to be enhanced. Currently it is called MSG-3. This standard for a specific type of aircraft is set up when the aircraft is being designed.

The remarkable output of this exercise was to identify the maintenance of significant items as well as required inspection items. The technical reality underpinning this process is that a mechanical item deteriorates in performance over time, and that there are a few significant items which directly affect the airworthiness of an aircraft. Hence as long as the operator keeps track of the performance of these items and maintains them proactively, the aircraft shall remain airworthy. Moreover, some items should be constantly inspected to see if unpredictable events such as change of wind pressures, lightning strikes, etc., have caused any damage, which would require maintenance.

It needs to be noted that aircraft were flying well before the widespread use of computers to support maintenance processes, and this is why the processes and methodologies that were developed to maintain an aircraft were mostly paper based and manual. Even today it is possible to achieve airworthiness without using information technology provided the number of aircraft being maintained is small.

It was not so very long ago, before deregulation, that the cost of maintaining the airworthiness of an aircraft was irrelevant. Airworthiness was maintained irrespective of the cost incurred, but with the increase in the volume of aircraft worldwide, and the increasing deregulation of the industry, the pressures to increase cost effectiveness has made it essential for operators to look more closely at using information technology effectively.

In this book we are going to talk about the aircraft maintenance in detail. Meanwhile, in order to set the context, it is worthwhile to take in a very high-level view of how it works.

An aircraft is designed to fly, exploiting the laws of physics and engineered to keep on flying for a significant period of time. However, each flight cycle, take off and landing leads to deterioration of the aircraft's ability to fly due to the impact of the environment, structural fatigue and operational incidents. The purpose of maintenance is to compensate for the debilitation of the aircraft due to those factors and bring it back, as close as possible, to its original ability to fly. The discussion on the theory of maintenance management is beyond the scope of this book. However, it is important to highlight that the requirement for an aircraft to safely fly passengers renders the method of maintaining it unique, which generally deviates from the theory and practice of general plant and equipment maintenance.

Let us elaborate this further. In qualitative and quantitative terms, the ability to fly safely, i.e. being airworthy, is defined as serviceability of all the airworthy significant components. In aviation maintenance parlance, it is that the components, which are essential for keeping the aircraft up in the air, are as designed and serviceable. The aircraft is then deemed to be airworthy. These involved components are called Maintenance Significant Items (MSI). We will also talk about the Structural Significant Items (SSI) later in the book. But to get a basic understanding of concept and its correlation with information technology, let us keep it confined to MSI.

In order to identify and categorise the significant items essential for airworthiness and the items that would need monitoring and maintenance, logic was developed by MSG-3. This is also known as MSG-3 logic in the industry. Each aircraft type is analysed using MSG-3 logic, thus arriving at a design reference point, which becomes the target for the achievement for each maintenance activity.

Each maintenance-significant item has a threshold of operation, either in flying hours or landings or cycles or any other parameter, after which the performance is deemed to deteriorate, impacting the airworthiness of the aircraft. Any significant item which is declared unserviceable impacts the airworthiness of the aircraft.

Maintenance is an act of making an unserviceable item serviceable, thereby improving the airworthiness of an aircraft. With thousands of unique parts in each aircraft, it is apparent that a significant amount of information needs to be managed to keep an aircraft airworthy cost effectively. As noted by the industry:

> The true definition of the word 'Airworthy' was never included in the Code of Federal Regulations until the 14 CFR Part 3, General Requirements, was established. The definition was included in the guidance, such as Advisory Circulars and Orders, but never in the Rule. Part 3 defines the definition of Airworthy as; the aircraft conforms to its type design and is in a condition for safe flight (*http://en.wikipedia.org/wiki/Airworthiness*)

Hence, it is the overall accumulated results of several reports and notifications, which are accepted by the regulating authorities as parameters for 'serviceability', that defines the airworthiness of an aircraft.

This book is about using the power of information technology not just to collect, disseminate and present the information to the regulating authorities but mainly to gain efficiencies and ensure the safety of passengers. The key is to achieve airworthiness efficiently and cost effectively for an aircraft so the passengers can safely travel though the air again and again.

I believe information technology has a lot to contribute towards achieving this airworthiness.

Note

1. Chandogya Upanishad, *The Upanishads*. Penguin Books. Translation from Sanskrit by Juan Mascaró, 1965.

An overview of aircraft maintenance

Abstract: This chapter highlights the criticality and immediacy of aircraft maintenance and elaborates the key strategies and objectives of the work at hand in order to set a context for aligning information technology provisioning to meet these objectives. It also familiarises the reader with some of the key concepts of aircraft maintenance.

Key words: aircraft maintenance, Maintenance Repair and Overhaul, MRO.

Then Bhargava Vaidarbhi asked: Master, what are the powers that keep the union of a being, how many keep burning the lamps of life, and which amongst them is supreme?[1]

In the world of commercial flight, it is often said that an airline's Flight Operations Department, whose job it is to plan the routing and scheduling of flights, would like to keep an airplane flying all the time; either flying between airports or taxiing; either preparing for take off or after landing; or, by necessity, being a commercial entity, loading and unloading passengers or cargo. On the other hand, the Engineering Department, whose job is to keep an aircraft safe and serviceable, wants to keep the aircraft on the ground, for maintenance, for as long as possible.

This rivalry is not limited to the aircraft's position in the air or on the ground. It also pervades each other's view on the justification for each other's existence. The ultimate dream aircraft for engineers would be the one which does not need pilots and the gnomes of the nether world who keep scheduling and rescheduling the flights just to annoy them.

On the face of it, the objectives of both the departments would seem to be diametrically opposite. But the irony is that the worst nightmare of an Engineering Department is when an aircraft is on the ground, unscheduled. A simmering alert goes out for AOG (Aircraft on Ground) when that happens. During the AOG, one who has witnessed the panic and controlled frenzy of aircraft engineers will confirm that engineers would also like to see the aircraft flying. In fact the main objective of the engineers is to keep the aircraft 'airworthy' and flying, for as long as possible.

Once I tried to compare this scenario with ships and trains. But then, it so happens that the trains remain constantly on the ground and ships do not cause panic when they are berthed, tethered to solid land.

In this chapter we will discuss why aircraft maintenance is unique and what guides the maintenance processes and its outcomes.

An aircraft undergoes a maintenance process when it is deemed to have lost its airworthiness. This could be immediate or in the future. For example, the AOG alert described above would be an immediate and unplanned activity that must be undertaken urgently or it will throw into disarray all the hard work of the Flight Ops Department, to say nothing of the inconvenience to passengers and loss of revenue for the airline. A scheduled maintenance event on the other hand is based upon a variety of factors and refers to an on-going and planned process for the life of the aircraft to maintain continuous airworthiness. It is this maintenance process that will be discussed in greater detail in Chapter 5.

1.1 Aircraft maintenance process

It is important to note, at the outset, that the main difference between general plant and machinery maintenance and aircraft maintenance is that aircraft maintenance is mandated and monitored by regulatory authorities, such as the FAA, CAA, etc., and therefore the aircraft maintenance process is highly standardised and enshrined in MSG-3 (Maintenance Systems Group) and other directives such as AC120-16D and JAR 145. These standards will be discussed later, but not in detail.

Whenever a new aircraft or aircraft type is designed, an MRB (Maintenance Review Board) is formed. This organisation comprises of regulatory authorities, OEM (Original Equipment Manufacturers) and other interested parties. Figure 1.1 shows in brief how a maintenance program for an aircraft is developed. This process is very well described in various ATA (Air Transport Association of America) documents.

Figure 1.1   Aircraft maintenance process

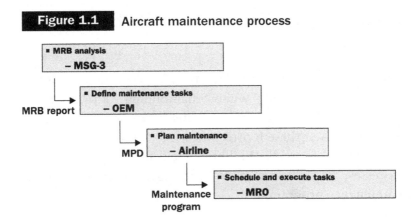

It is no surprise that the regulatory compliance lies at the heart of aircraft maintenance, hence the process is prescriptive. This means that no matter where aircraft maintenance is performed, whether San Francisco or Timbuktu, the process remains the same for the same aircraft type. In principle, there are no two ways of maintaining an aircraft. The maintenance organisation conforms to the directives of the aircraft manufacturer and regulatory authorities. Maintenance staff are trained exactly the same way across the world and their tasks are also defined uniformly.

The MSG-3[2] refers to the recommended aircraft maintenance process as 'Task Oriented', meaning that the entire sets of tasks are defined from beginning to end. Thus an engineer cannot carry out a task on an aircraft without defining and getting the task approved beforehand, just as it is a measure of the completeness of the task description in the MSG-3 procedure that a Licensed Aircraft Engineer cannot turn the screwdriver three times if the Task Card says twice.

The tasks, in the 'Task Oriented' maintenance, are grouped as follows:

- airframe system tasks
- structural items tasks
- zonal tasks.

Altogether, there are fourteen identified core unique tasks that are performed while maintaining an aircraft. Maintenance activities are essentially permutations and combinations of these fourteen tasks. A combination of one or more tasks forms the building blocks for the maintenance processes, and one or more such processes will form a maintenance program.

These tasks are like the 'roots' of a language, such as Sanskrit or Latin, which generate lots of words. Similarly these tasks help create a number of maintenance activities, which in turn create work packages, which become the building blocks of maintenance programs.

MSG-3 is a process, which is used by the MRB to define and categorise maintenance activities based on the logic of 'consequence of failure' of aircraft components, for an aircraft type. In other words, it defines which components *must* be maintained, and how, to ensure that this aircraft type is airworthy. The OEMs use the outcome of this to provide the operators with a recommended maintenance plan, which is called MPD (Maintenance Planning Documents/Database). The operators create their own maintenance plans based on the MPD and subsequently schedule their maintenance operations. Thus, an aircraft is maintained using approved maintenance programs.

The entire philosophy of aircraft maintenance is based on the MSG-3 failure logic. However, detailed discussion on this topic is beyond the scope of this book.

One must keep in mind that almost all of the maintenance activities are preventive in nature, from an airworthiness perspective. The concept of 'break–fix', which is common in general plant and machinery maintenance, is not applicable to aircraft. As we can imagine, there is no maintenance required after an aircraft has fallen out of the sky.

Aircraft maintenance is conducted in two modes: scheduled and unscheduled. There is a very popular notion within aviation engineers, who plan and schedule maintenance programs, that the need for unscheduled maintenance should never arise. However, in the real world, this is far from true. Things happen, and unscheduled maintenance has to be undertaken.

Unscheduled maintenance is a subset of scheduled maintenance in the regimented world of aircraft maintenance. Hence, most of the discussions in this book will centre on scheduled maintenance. In fact, from an information technology (IT) perspective, a system which can support scheduled maintenance can also support unscheduled maintenance.

For a process that is so clearly and unambiguously defined, and where procedures and frequencies of activities are legally mandated, monitored and are globally consistent, it would seem incredible that IT has not been applied with force to make the process of reporting and execution of maintenance tasks highly efficient and cost effective. But, unfortunately aviation MRO has yet to fully leverage IT.

While increasing efficiencies have been brought to all aspects of commercial aviation as a result of the power of information technology, such as flight booking systems, cargo handling systems and increasingly sophisticated baggage handling systems using radio-frequency identification (RFID) tags to mention just a few, the area of aircraft maintenance appears to receive scant attention and investment from the IT industry. In fact, most aircraft maintenance systems in use today are run on mainframes and were developed in the 1970s. This topic is further elaborated in Chapter 4.

Compounding the lack of new and acceptable solutions from the IT industry is the general apathy of the aircraft engineers in adopting the available information technologies in the MRO space. This apathy should not be taken as resistance to change and innovation, as these engineers, in common with their peers in the other technology literate domains, are often keen consumers of the latest in computers and electronic gear. Rather, this behaviour can be attributed to two aspects of the practice of aircraft maintenance: first, the processes are so prescriptive and so effective that the practitioners attain high levels of skill in relatively short time. Second, it is common that the engineers are more inclined towards enhancing their engineering skills than improving their IT skills. Though IT promises to make the activities transparent and more productive, the current deployments of IT systems has resulted in increased complexity of operations in many instances.

Finally, and perhaps most importantly, modern commercial aviation came to maturity prior to the maturity of modern business support systems from the IT industry, leaving a gap that is still apparent in the area of MRO.

When the Boeing 747 was introduced in the 1970s, IT was still in its infancy and therefore the maintenance processes were designed to be paper based. The engineers learned to keep the sophisticated aircraft flying without the need for a computer-based system. To this day, an engineering department does not technically need a computer-based system if the number of aircraft to be supported is small.

Normally, the use of a computer system is dictated by volume and not the complexity of the process. This is why even today a significant number of aircraft maintenance programs are managed using a PC, or at best a small server.

Later in this book we will discuss the objectives, strategies and organisational and compliance requirements for aircraft maintenance in greater detail, but for the purpose of simplicity we will use the terms Engineering and Maintenance (E&M), Maintenance and Engineering (M&E) and MRO interchangeably.

1.2 Maintenance objectives

It goes without saying that the main objective of aircraft maintenance is to keep the aircraft flying for as long as possible. In light of this, MSG-3 has defined four measurable objectives.

The objectives of efficient aircraft scheduled maintenance are:

1. To ensure realisation of the inherent safety and reliability levels of the aircraft.

2. To restore safety and reliability to their inherent levels when deterioration has occurred.

3. To obtain the information necessary for design improvement of those items whose inherent reliability proves inadequate.

4. To accomplish these goals at a minimum total cost, including maintenance costs and the costs of resulting failures.

These objectives recognise that scheduled maintenance, as such, cannot correct deficiencies in the inherent safety and reliability levels of the aircraft. The scheduled maintenance can only prevent deterioration of such inherent levels. If the inherent levels are found to be unsatisfactory, design modification is necessary to obtain improvement.

Since such design modification is almost never undertaken by a commercial aviation organisation, but rather by the aircraft manufacturer, the measures of the maintenance objectives are 'Despatch reliability' and 'Turnaround time'; and the objectives can be summarised as: 'Prompt delivery of airworthy aircraft to the airline reliably, consistently and cost effectively'.

From an engineer's perspective, the first three objectives are relevant, clearly measurable and consistently achievable. It is the fourth objective, which creates the greatest angst among the engineering community.

Engineers do not like 'bean counting' and are a proud bunch, who keep 100+ tonne machines up in the air. The cost of keeping this airworthiness is something which is very difficult to manage. It is not just because engineers do not like accounting, but the fact that MSG-3 or any other regulatory directive only specifies some general statements about managing costs. There are no standard Key Performance Indicators (KPIs) or processes defined to track and optimise maintenance costs. The regulatory authorities have left this to the 'operators' internal accounting practices and processes. Strangely enough the rigour that we see on the engineering side is glaringly missing in the financial aspects.

To add to this conundrum, the maintenance processes, per se, are heavily inclined towards airworthiness delivered by labour only. MSG-3 takes labour into account but more or less ignores materials and logistics costs. Hence there is a lack of fiscal discipline within the practice of aircraft maintenance. Having said that, there are a significant number of aircraft operators who have stitched together costing methods and practices, but there are as many different models as there are operatives.

It is in this space that IT can, and in some cases does, play a major role in achieving the objective of managing costs.

1.3 Aircraft maintenance strategies

The main challenge for aircraft maintenance engineers is to keep the aircraft airworthy without any service interruptions. Hence, in theory, an aircraft should visit a hangar only as planned. It must be kept in mind that hardly any maintenance can be done while the aircraft is flying. A maintenance engineer on a flight, if present, is not allowed to carry even a screwdriver! The other aspect of safety linked with security.

So how, in fact, is an aircraft to be maintained? As the objectives are mandated by the regulatory authority and the processes defined by MSG-3, the industry has adopted the following strategies:

- *Redundancy:* This is built into the aircraft systems and components, by design, so that failure of one system or component does not affect as aircraft's airworthiness. The maintenance on these items can be performed at the best opportunity available. Some modern aircraft have up to seven levels of redundancies on certain critical components. As is apparent, this strategy increases the initial cost of the aircraft significantly and adds to the unused inventory for the operators. Moreover, this strategy can apply to a limited number of systems and components such as radio, radar, etc.

- *Line replaceable units (LRU):* The LRUs are designed modular components, which can be replaced very quickly when they go faulty. This implies that the failure of the module does not impact an aircraft's flying abilities. In addition to that, these modules can be monitored for symptoms of a fault occurring. This strategy allows the maintenance activity to take place while the aircraft is in service. It does not mean while it is flying. It means when the aircraft is still docked to the airway bridge or being prepared to turn around for another flight. Hence, the Line Maintenance Engineer does not have to repair the component on the spot, but just replace the faulty one with the serviceable one and send the unserviceable component to the workshops for repair. This strategy is being used more and more in the industry as the new aircrafts are designed around the modular concepts.

- *Minimum equipment list (MEL):* This is a list of equipment which must remain serviceable for the aircraft to be airworthy. In other words, not all components need to be serviceable to keep the aircraft safe for flying. The strategy of maintaining this list helps focus the maintenance organisation to plan and execute defect rectifications in a much more focused way. This list is derived from the Master MEL from the OEM; and approved by the local regulatory authorities.

A combination of the above strategies allows an operator to keep the aircraft flying, with minimum unplanned interruptions.

1.4 The maintenance organisations

Normally the structure of an organisation follows its strategy. However, in the case of aircraft maintenance the core structure of the organisation is mandated by the regulatory authorities. This does not imply that all the aircraft maintenance organisations will look the same, but the core departments must be organised as mandated by the FAA in its document AC 120-16D and other similar directives by other aviation regulators.

In the early days of commercial aviation, aircraft maintenance was mainly carried out by the engineering department of an airline. But as the volume of work increased, airlines identified aircraft maintenance as a non-core function. Hence, independent organisations have come into being who undertake aircraft maintenance for a number of airlines, effectively operating as outsourcing entities.

At first the large airlines started the trend of hiving off their engineering departments as either wholly owned subsidiaries or independent companies, and later smaller airlines followed suit. This trend is continuing and it is conceivable that in the not too distant future, none of the major airlines will have captive engineering and maintenance divisions.

All these new entities now call themselves aviation MRO organisations. They either operate as profit centres or independent companies. In affect they still mainly cater to their mother airline, but most strive to add to their revenue by undertaking work for other airlines. This trend has helped the LCC (low cost carriers) the most.

In addition to these organisations there are independent aviation MRO organisations, which do not have any affiliation to any specific airlines. There are a number of joint ventures coming up as well.

From another perspective these organisations are also capability specific. Some maintain just the engine or a specific type of aircraft or specific type of maintenance, e.g. line maintenance or structural maintenance, etc.

The information requirements of all these different types of aviation MRO organisations are different and will be discussed in detail later in the book.

1.5 Regulatory compliance

There are many similar standards and guidance in the maintenance industry, which regulate the maintenance processes. However, in the aviation MRO industry the emphasis on discipline and rigour is much more pronounced than any other type of maintenance.

It is fine to see that the aircraft OEM and MRB specifies and even mandates the maintenance processes, down to the task level. However, it should be an absolute delight to the flying public that these standards and procedures are religiously adhered to. This is how it is done.

In summary, all the sovereign nations, barring a few, have agreed to a governance process for ensuring the safety of the flying passengers. Each one of them has legislated aviation laws, which are patterned after the US FAR (Federal Aviation Regulations). They also establish government funded Civil Aviation Authorities, which get affiliated with the regional regulatory associations and authorities. They all have similar organisational structures and mandates for operation. This book will not discuss the Civil Aviation Authorities and their modus operandi. There is copious literature on this topic. The readers are recommended to look up some of them.

So how do national authorities contribute towards the airworthiness of an aircraft? It works like this: an airline can operate an aircraft only if it has been registered by the national Civil Aviation Authority. The process of registration includes verification and agreement of the maintenance processes to be undertaken by the airline. Once satisfied, the Civil Aviation Authority issues a registration number (known as a *tail number* in the industry), which is prominently displayed on the aircraft. These tail numbers remain valid while the aircraft is in service and the operator remains in the same geographical region or country.

Thenceforth, it is mandatory to log all the maintenance activities on the aircraft and report regularly and on demand to the Civil Aviation Authority.

This Civil Aviation Authority has the authority to certify whether the aircraft is airworthy or not. However, it also has the authority to delegate to a specific person or organisation the authority to certify.

Hence, the maintenance regime that is agreed in the MRB report, based on the MSG-3 logic, is maintained by the Civil Aviation Authorities.

1.6 Conclusion

In spite of the fact that the records can be maintained manually, the volume of data and the complexity of the maintenance process lends itself very well to leveraging IT. Besides the regulatory reporting requirements and maintenance of data lineage, with history, IT can be a fabulous tool for increasing efficiencies and cost effectiveness of the aviation maintenance processes.

1.7 Notes

1. Prasna Upanishad, *The Upanishads*, Penguin Books. Translation from Sanskrit by Juan Mascaró; 1965.
2. ATA MSG-3; Operator/Manufacturer Scheduled Maintenance Development; Revision 2003.1; Copyright © 2003 Air Transport Association of America, Inc.

2

The business of maintaining aircraft

Abstract: This chapter highlights the business processes used to maintain aircraft. The chapter provides an overview of demand and capacity planning; describes the service offerings; highlights types of customers and their contracts; identifies enablers to significantly increase the use of information technology in order fulfilment; and describes some of the key areas such as warranty and customer experience. None of these critical business activities are specified by any regulatory authorities nor is there an industry standard for best practice.

Key words: MRO, regulation, FAA, CRM, MSG-2, MSG-3, contract, demand, order.

He who knows all and sees all, and whose glory the universe shows, dwells as the Spirit of the divine city of Brahman in the region of the human heart. He becomes mind and drives on the body of life, draws power from food and finds peace in the heart. There the wise find him as joy and light and life eternal.[1]

2.1 The aviation MRO market overview

Maintaining aircraft is potentially a money making business. It is very similar to the situation where if you maintain your own car it is a cost to you but if you take it to a garage it is revenue to the garage owner. Initially, all the airlines had (some still do) their own maintenance facilities and called them M&E department or divisions. These Maintenance and Engineering departments were cost or burden centres. In other words the airlines treated them as necessary evils. And devils they were, as long as the regulated skies lasted. The M&E divisions could blackmail the airline management by levelling the AOG guns at their masters' heads and the airline would pay any amount demanded in the name of 'airworthiness' and aircraft availability. We will discuss the impact of this behaviour on IT investments by M&E later in the book.

Meanwhile, two phenomena happened: de-regulation of skies and outsourcing. De-regulation debilitated the airlines to an extent that they could stand up to the M&E divisions and refuse to pay the ransom demanded. The power emanated out of being forced to cut costs as prices crashed. In addition to that, their ability to outsource gave airlines the power to threaten and put down the arrogant engineers (no pun intended).

These M&E divisions started morphing into MRO organisations and found that they could really make money while maintaining aircraft like their brethren on the street corners maintaining automobiles.

It is predicted that by 2018 there will be plenty of aircraft to maintain. The number of aircraft flying across the globe by then is expected to be somewhere around 28 000. The global fleet is expected to increase by 9000 aircraft from

2007, a 4% compound annual growth rate with the installed base of 27 966 aircraft comprising 6128 future orders, 7314 firm orders, 206 parked aircraft and 14 318 aircraft in the retained fleet.

The Asian and European fleets in 2018 are expected to be nearly the same size as the North America base. In year 2010, orders stand at 6900 aircraft.

The 2017 MRO market is expected to be worth $61 billion, a $15 billion increase since 2007. Modifications will account for $5.3 billion (a 2.8% compound annual growth rate), heavy airframes $7.2 billion, and the component business – the highest compound annual growth rate at 3.9% – $13.3 billion.

Line maintenance work is valued at $13.1 billion, while engine MRO totals $22.4 billion of annual work, a 3.8% compound annual growth rate.

In total, air transport MRO will account for 45% of a broader $140 billion global maintenance market, with outsourcing at 75% – up from 52% 10 years previously, although this could yet be owned or partially owned by airlines.

So, how do they make money? We will discuss the business model generally adapted by an aviation MRO organisation in the following sections. The areas covered in the sections below are mostly beyond the purview of the regulatory authorities and do not necessarily have a standard best practice. However, they describe the money making part of the MRO business.

2.2 Customers – who are they?

Most of the MRO organisations have grown from being a M&E department of an airline. FAA in its documents generally assumes that maintenance of a commercial aircraft is done by a department or a division of an airline. It goes so far as to recommend an organisational structure for such an organisation, which falls short for an independent MRO organisation. Hence, an MRO organisation generally starts with a captive market, i.e. the mother airline. However, now the trend is for MRO organisations to be set up either independently or as joint ventures with airlines and OEMs.

The customers of an MRO organisation can be categorised as follows:

1. The mother airlines: typically if the organisation is linked with the mother airline then the majority of the business is concentrated there.

2. LCCs and small carriers: they typically outsource the maintenance work

3. Leasing companies: they use multiple MRO organisations

4. Airlines taking location advantage: they typically use the MRO organisations, which suit their aircraft routings.

5. Airlines looking for cost reduction on special services.

There are approximately 8000-plus registered airlines and about a dozen leasing agencies. Among them they need or will need maintenance on more that 25 000 aircraft.

These operators can be segmented as shown in Table 2.1.

Table 2.1   Airline segments

Type	Fleet size	Employees	Number
Tier 1 Airline	70+	3000+	25
Tier 2 Airline	12 to 70	200 to 3000	274
Tier 3 Airline	1 to 11	Up to 200	8000+

An MRO organisation typically targets four to five of these organisations as long-term accounts. There are rarely one-off customers but some overflow work from other MRO organisation also turns up. Some smaller MRO organisations take up subcontracting work from larger MRO organisations as well.

2.3 Demand and capacity planning

In line with any other type of business, MRO organisations conduct demand and capacity planning. Initially, at their inception, they were attached to their mother airlines, where the basic information for maintenance demand planning was easily available. However, once they became independent entities, the acquisition of data required for demand planning became challenging.

Despite the fact that demand planning and capacity planning are interrelated, let us discuss them separately first and then look into the interactions between them later.

I paraphrase the economic definition of demand: 'In aviation MRO, *demand* is the aircraft operators' desire to get the aircraft maintained; and the ability to pay for it as well as willingness to pay. The term demand signifies the ability or the willingness to buy a particular maintenance service at a given point of time.'[2]

At the operator's end, demand is triggered by two events:

1. Scheduled maintenance requirements.
2. Unscheduled maintenance requirements.

The scheduled maintenance requirements are generated by the MPD-based maintenance plan. This includes modifications, mandatory or otherwise. They are also dependent on the operation of the aircraft in terms of flights and routing. The MRO organisation may or may not have access to this information. Normally, the organisation will undertake a long-term contract which provides a good visibility of the operator's requirements for demand planning.

The unscheduled requirements are either results of incidents or mandatory modifications or start/end of lease or phase-in/phase-out of the aircraft.

An MRO organisation normally looks at its location and resource capabilities and assesses the demand. Currently, long-term demand planning is hardly ever done by the MRO industry. The demand is taken at a macro level and then various other parameters come into play.

There are many point solutions, not necessarily integrated with the core aviation maintenance solutions, which are available in the market. However, the IT industry has yet to come up with an offering that can solve the dual problem of forecasting the probability of failure rates and the affect of preventive maintenance on the actual failure rates or airworthiness of an aircraft.

2.4 Service offerings

Not all MRO organisations are the same. Their shape, size and characteristics are defined by the type and range of services they offer. These services are predefined and need to have been certified for a specific aircraft type. In the aviation MRO world the act of carrying out maintenance is known as a 'check' being conducted. You do not maintain an aircraft, you perform a check and these checks are predefined.

The services offered by a typical MRO organisation can be categorised as follows:

Heavy Maintenance

Line/Ramp Maintenance

Integrated Fleet Management

Other services.

(a) Heavy Maintenance, also called Hangar Maintenance, is performed, as the name implies, in a hangar. Another term used for these visits is 'Heavy Check'. These services are also known as the 'Lettered' Checks. This term was first defined in MSG-2 and they range from letters A to D. 'A' Check had the smallest work content and 'D' Check the largest. The latter involves almost completely stripping the aircraft and rebuilding it. In MSG-3 the lettered checks are not relevant because all the maintenance work is task oriented. However, the industry has still stuck to the MSG-2 terminology because MSG-3 does not offer a way to define the service offerings like MSG-2. It is easier to sell a 'C' Check than describe a full set of tasks, which are bundled but there is no standard way of naming them.

Therefore, to keep to the traditional way of defining the offerings, an aviation MRO organisation sells A, B, C and D Checks. With the advent of MSG-3 and the way the work is packaged, B checks are almost out of fashion and A Checks are mostly performed on the Ramps. All these checks fall under the category of scheduled maintenance and are generally planned by the operator or an MRO organisation on behalf of the operator. 'C' and 'D' Checks are also called Major Check because of the sheer size of the work content. An average 'C' Check could consist of more than twenty thousand tasks and have a turnaround time of three to four weeks. A 'D' Check could run for months. The maintenance planners use phasing of 'C' checks and call them '1C', '2C', 'C1', etc. But these are not granular enough to be defined as specific and discrete service offerings

(b) Line or Ramp Maintenance, as the name implies, is conducted while the aircraft is still in service. The tasks are mainly to fix the defects reported by the pilot or crew, or some of those, which were deferred. The tasks have to be completed within the turnaround time of the flight, which is normally less than an hour. These include visual checks and inspections. These services are mostly incidental and do not have any specific naming convention.

(c) Integrated Fleet Management is an umbrella service, which covers everything that is required to be done in terms of maintenance to keep the aircraft airworthy. Under this offering the operator hands over almost complete responsibility of aircraft maintenance to the MRO organisation. Hence, the MRO organisation ends up influencing the routing of the flights and flight schedules of the operators. Sometimes a grey area remains as to how far the MRO organisation should go into the realm of flight operations. Currently, very few software claim to support this kind of service but it is expected that the demand will force the software vendors to come up with integrated solutions.

(d) The other services that an aviation MRO organisation provides can range from maintenance planning to carrying out a specific campaign-based service, like painting the aircraft to show the operators' participation in or sponsorship of events like the Olympics, etc. Most of these offerings are not directly related to the maintenance work per se.

As of now these service offerings are not very standardised in the industry. Hence the MRO organisations have to undergo a complex series of steps to define and manage a contract with the operators. This activity is elaborated in the following section.

2.5 Customer orders and contracts

An integrated MRO contract can include airframe/ component maintenance, rotable logistics and/or fleet management options, which require key strengths in aircraft health management systems and distribution networks. However, an MRO organisation signs up various types of contracts for its service offerings.

Normally an integrated contract is based on some fixed terms such as flying hours or fleet type and number, etc. They can also be based on an overall capped expense budget. These types of contracts are often long term, five to ten years.

Similarly, the contracts could be for all the major checks and based on actual time and material spent. This is a very common type of contract but this is also a major source of disputes between the operators and the MRO organisations. The issue created by this aspect of the contracts are further discussed in Section 2.7 later.

In the aviation MRO business there are very few instances where operators place incidental orders. However, this is very common between MROs themselves. Normally an MRO organisation will have a contract with the operators as well as other MRO organisations. But in most cases the arrangement with other MRO organisation is an understanding that is invoked as and when a need arises for a specific type of service.

There are significant amounts of paperwork involved in the contract and order management. In addition to that the service order cycle does not follow any specific method or process as in the automobile industry. In spite of the fact that the incidental orders are very rare, the contract management is still not standardised. Hence, there are very few software which cater to this aspect of the business. The order cycle is discussed in further details in the following section.

2.6 Order commitment and fulfilment

In the aviation MRO world, it is very difficult to commit an order with a high degree of confidence as the volume of work and related schedule for an order is very difficult to estimate and predict. This is because the unscheduled component of the work is normally very high. An aircraft can come with a planned work package but when the actual work starts the volume can exceed more than a hundred per cent. A prudent organisation would normally factor this in but this does not solve its logistics problem that arises due to the unknowns. Some of the 'unknown' tasks may require spare parts, which have very long procurement cycles or may not be available in time or may not be available at all.

The MRO organisation, therefore, ends up having to put a significant amount of contingencies into their estimates. this does cover the risk of commitment but it does widen the gap for dispute with the operators.

When an MRO organisation commits to an order, in terms of price and schedule, it tends to carry a lot of buffer, which encourages the operators to try and squeeze the price down. The operators always feel that they are being overcharged. The trouble is that the MRO organisations contribute to that feeling by sometimes undercutting each other or agreeing to unreasonable schedules.

2.7 Invoicing

A common business practice is that an invoice is raised by the performing organisation at specific milestones contractually agreed, and at the completion of the job. However, in a typical MRO job the complexity of cost collection results in the final invoice being sent much later and after the aircraft has left the hangar. It is not unusual to see most of the invoices ending in dispute.

> An *invoice* or *bill* is a commercial document issued by a seller to the buyer, indicating the products, quantities, and agreed prices for products or services the seller has provided the buyer. An invoice indicates the buyer must pay the seller, according to the payment terms. The buyer has a maximum amount of days to pay these goods and are sometimes offered a discount if paid before.[3]

In reality the nature of the MRO work is such that it is impossible to put down a fixed number for a work package. There always is a big difference in an order commitment and its fulfilment as discussed earlier.

Since the MRO organisations did not have to bill their mother airlines, their major customers, they are not generally geared for prompt and accurate invoicing.

The difficulty of cost collection is inherent in the way the job is carried out to maintain the aircraft. This is compounded by the different contractual and warranty-related issues that come into play.

Again the process of invoicing in the MRO world is still evolving and the standard software packages have difficulty in supporting this function.

2.8 Warranty in the world of aircraft maintenance

Some definitions of warranty in general:

> *A guarantee given to the purchaser by a company stating that a product is reliable and free from known defects and that the seller will, without charge, repair or replace defective parts within a given time limit and under certain conditions.*[4]

> *An offer, often associated with a purchase, in which a marketer provides customers a level of protection, beyond a Guarantee period, that covers repair or replacement of certain product components if found defective within some identified time frame.*[5]

An aircraft is delivered to an operator with a considerable amount of warranty in terms of dollar value and complexity. The warranties are provided by the airframe builders as well as other OEMs (Original Equipment Manufacturers). In addition, the MRO organisations provided their own warranty on the components they repair, service or overhaul, sometimes over and above the warranty provided by the OEM or after the OEM warranty has expired. All these are recorded in warranty certificates, which form the legal commitment from the warranty provider.

A significant amount of revenue is earned by an MRO organisation by managing the warranties applicable to an aircraft undergoing maintenance. However, a warranty on an aircraft, for an MRO organisation, works in two ways. The first is where it collects and passes on the warranty benefits from the OEMs to the airlines. The second is where the MRO organisation provides a warranty to the operators for the work done. Obviously the money is made on the first type. The second one creates liability for the MRO organisation. Both processes are complex and require enormous amounts of information gathering and tracking.

Let us first discuss the collection and consumption of warranty benefits from OEMs.

Each OEM provides warranty coverage on the items that it supplies, e.g. airframe, engine, landing gear, radar or even a simple pump. This type of warranty is also called 'in-bound' warranty, where the liability for the warranty is carried by the OEMs. Hence, it is not uncommon to see OEMs reminding the operators if the rate of warranty consumption is less than expected. The driver for this behaviour is that the OEM providing the warranty wants to reduce its liability, similar to what employers do when they ask their employees to use their annual leave to reduce the liability carried by the employers.

All the warranties, which are mostly specified in a document, are identified and recorded by the operators, irrespective of where it originates: OEMs or MRO organisations. Warranty information, which is received in an unstructured form, is electronically captured and stored in a structured form. This is to facilitate recording and tracking of warranty consumption, validity and limitations or constraints. Almost all warranties have multiple conditions and multiple levels of constraints. The MRO organisations collect this information from the operators and use it for warranty processing.

Some of the claimable warranties are known to the MRO organisation before the aircraft rolls into the hangar. This happens if the operator has diligently passed on the list of defects under warranty with failed parts identified. However, most of the issues are discovered as the maintenance process moves into execute mode. You open a panel and there you go, a failed or unserviceable part is found and it could be under warranty.

A warranty is claimable for a failure of the component or an aggregate within a limit, such as flying hours or age; or it can also be based on degradation of performance. There is good potential for revenue in these warranties for the MRO organisation, because most of the warranties cover labour and material costs for removal and fitment of the failed part. In such cases the MRO organisation recovers the money for labour and additional material from the OEMs.

Once the failed or unserviceable part is identified, the MRO organisation wants to identify whether the part carries a warranty or not. If yes, then the warranty is acted upon. The part is either shipped to the repair agency specified in the warranty or if the MRO organisation is authorised, the repair is carried out by the organisation itself. In either case the cost of removal, packaging, shipment or repair is recovered by the organisation.

In the case of 'out-bound' warranty, the MRO organisation prepares a warranty document and sends it to the operator along with the repaired part. Whenever a part covered by this warranty appears on its doorstep the organisation is obliged to repair the part free of cost or compensate the operator if the repair on the same part is carried out by a third party. Sometimes there could be a residual warranty on a part, which needs to be managed. In other words, the cost recovery becomes a very complex and tedious exercise.

As we noticed, the warranties are generated by two types of organisations: OEM and MRO. However, the management of these extends to operators, who are the direct beneficiaries of the warranties. MRO organisations are the indirect beneficiaries as well as providers. Hence, the extent of warranty management becomes more complex for them. Their warranty processes have to encompass both the way an OEM manages and the way an operator manages. They earn as well as pay for warranty, depending on the situation described earlier.

The warranty management process in the aviation industry is not yet fully standardised and all the involved parties seem to have warranty management systems of their own, although there are commonly accepted best practices.

Some of the applications currently in use follow some of these best practices. This aspect of aviation MRO is not regulated and is not linked with 'airworthiness' requirements. However, from a revenue perspective a good IT solution can create significant value.

At a very high level the warranty provider requires an application to:

- Create warranty documents and related information in structured format.
- Communicate this information electronically to the operator.
- Track warranty consumption and trigger parts procurement.
- Link up with their own financial and inventory management applications.
- Link up with the operator and MRO organisations' applications.

An aircraft operator requires an application to:

- Link up with the OEM and MRO organisations' applications.
- Receive warranty documents and structured information.
- Communicate with MRO organisations to send and receive warranty consumption information.

- Link up with their own financial and inventory management applications.
- Track and manage warranty consumption and trigger related activities.

An aircraft MRO organisation requires an application to:

- Link up with the operator, OEM and MRO organisations' applications.
- Receive warranty documents and structured information.
- Communicate with MRO organisations to send and receive warranty consumption information.
- Link up with their own financial and inventory management applications.
- Track and manage warranty consumption and trigger related activities.

2.9 Customer care

An aircraft operator expects an MRO organisation to provide a reliable customer service. However, this aspect of service is not very mature in the aviation MRO world. Since the main focus is on airworthiness, the softer aspect of the customer service is often ignored.

The lack of customer service in the industry could be because once the aircraft is out of the hangar the MRO organisation literally loses sight of the customer's needs. However, with the integrated fleet management now becoming more popular, it has become imperative for the MRO organisations to provide customer care, which means proactively engaging with the operators in supporting their aircrafts. This could be at times over and above the issues related to airworthiness.

In order to successfully provide customer care, the MRO organisation needs to keep more than technical information of the aircraft. This involves keeping tabs on the stakeholders within the operator's organisation. In other words, it requires Customer Relationship Management (CRM). Currently, very few MRO organisations fully use CRM software and also most of the CRM software does not fully cater to aviation MRO organisations' needs.

2.10 Conclusion

The final act of delivering an airworthy or serviceable aircraft to the customer is making the aircraft available for flight to the Flight Operations department of the airline. This exercise is known as the 'Tail Assignment' of the aircraft to the airline's flight schedule. The 'Tail', in this case, represents the registration number of the aircraft, as provided by Civil Aviation Authorities. You can see this number on the tail of any commercial aircraft.

The MRO organisation makes an aircraft available to be operated on designate flights for which it is paid for.

2.11 Notes

1. Mundaka Upanishad, *The Upanishads*, Penguin Books. Translation from Sanskrit by Juan Mascaró, 1965.
2. *http://en.wikipedia.org/wiki/Demand_(economics)*
3. *http://en.wikipedia.org/wiki/Invoice*
4. *http://www.thefreedictionary.com/warranty*
5. *http://www.knowthis.com/marketing-terms-and-definitions/marketing-terms-letter-w/*

Aircraft maintenance paradigm

Abstract: This chapter highlights the maintenance processes used to maintain commercial aircraft and provides a detailed breakdown of the maintenance processes. It describes issues and inherent challenges; highlights commonly used maintenance practices; identifies enablers to improve the effectiveness in the maintenance approach to aircraft maintenance; and tries to tease out some of the promising uses of information technology for airworthiness.

Key words: APQC, AMTOSS, SAP, process model, IBM, IPC, Task Card.

Intention is clearly greater than the mind, for it is only after a man has formed an intention that he makes up his mind; after that, he vocalizes his speech—and he vocalizes it to articulate a name.[1]

3.1 The paradigm

Intellectual perception or view, accepted by an individual, or a society; as a clear example, model, or pattern of how things work in the world. This term was used first by the US science fiction historian Thomas Kuhn (1922–96) in his 1962 book *The Structure Of Scientific Revolution* to refer to theoretical frameworks within which all scientific thinking and practices operate.[2]

In the previous chapter, we discussed the 'business' of maintaining commercial aircraft. In particular, how the aviation MRO business generates revenue with a goal to being profitable. That was a business venture point of view.

In this chapter, we are going to discuss how an aircraft is maintained. This will be a business process point of view. To use terminology familiar to IT Enterprise Architecture, the subsequent sections could be interpreted as a business architecture view; however, I am taking a more open approach, as this book is not intended as a standard for IT in the world of aircraft maintenance.

It is possible to map the processes discussed below to an industry standard business process model, if it exists. To my knowledge, there is no official standard MRO process model under the American Productivity and Quality Center (APQC) or similar standards organisations. Nevertheless, it is worth noting here that there are two major IT players who claim to have such industry reference models. One of them is SAP and the other is IBM. But these models are not in the public domain, they are proprietary, and therefore will not be discussed in any detail in this chapter. However, in order to get a proper perspective of the models, it is necessary to describe these at a very superficial level.

The SAP application suite comes with an industry reference model as part of its MRO package. In other words, if you buy and implement the SAP MRO solution then its reference model is available to you as a paying customer. The process model is not applicable to any other solution. Therefore, I would not call it a reference model but a MRO solution implementation methodology for SAP.

IBM's MRO process is sold independently of any software and IBM claims that it can be used as a template for any MRO solution implementation. This was developed around 2001 by PwC (PricewaterhouseCoopers), before IBM acquired them; and was based on their experience in implementing SAP for MRO at British Airways and a few other airlines. The original idea was to embed this model into the SAP implementation methodology. However, due to contractual constraints this did not happen. This model, while sold independently, is not yet approved by APQC[3] or any other industry standards organisation, and therefore cannot be considered to be an open MRO industry process model.

The irony is that the process of aircraft maintenance described in the following sections, in all seriousness, does not go very far from either of the models, because it cannot. Whatever you do in terms of aircraft maintenance is dictated by regulatory compliance, and therefore the process remains, and has remained the same since the advent of the Boeing 747 and it could well be that it will remain the same for some time to come. However, the industry needs to look at adopting a 'standard' reference model, which could enable the standardisation of the architecture and building blocks of aviation MRO software.

Maintaining a commercial aircraft, especially a passenger carrying aircraft, requires far more attention to safety related regulations and governance than any other type of aircraft. However, in the discussions in this book, we will not differentiate between types of commercial aircraft.

The aviation maintenance processes, on the surface, seem very similar to the general plant and machinery maintenance processes. However, the regulatory processes designed to keep the aircraft airworthy makes aviation MRO quite different.

In the following sections, we will discuss various life cycles – starting with the aircraft life cycle itself. It is essential to understand this life cycle before diving into the component maintenance life cycles, which will include the airframe, engine, component and GSE life cycle. These are the four domains where different types of maintenance strategies are used. I hereby take liberty of misquoting the famous definition of an aircraft by Boeing:

> *An aircraft is nothing but thousands of components and multiple engines, fitted to an airframe, flying in unison; which need to be maintained using Ground Support Equipment so that the aircraft can fly again and again till it is retired or in some unfortunate circumstances rendered incapable of flying.*

Let us now embark on the journey of analysing the various life cycles to understand and then create a framework for an IT architecture, which will not only support but enable the business of maintaining an aircraft to become more and more efficient.

3.2 Life cycle of a commercial aircraft

> *Life Cycle: The total phases through which an item passes from the time it is initially developed until the time it is either consumed in use or disposed of as being excess to all known materiel requirements.*[4]

The desire to fly was always present in humans since the time they saw the birds fly. It took millions of years before we could even articulate how this could be achieved. However, all the early civilisations dreamt of this ability and put it in their legends, myths and stories. Nearly all ancient cultures contain myths about flying deities.[5] Some thought of winged deities like Ahura Mazda of Persia; some without wings like the Hindu deity Hanuman. Similarly various flying vehicles were also dreamt of. There was the imaginative 'Pushpak', a flying chariot mentioned in the Ramayana, and the almost engineered famous flying machines of Leonardo da Vinci.

The Wright Brothers finally gave a real flying machine to the world. This transition from the imagination to reality took a very long time. But once this happened there was no stopping the human race, which achieved its objective of flying within less than a hundred years.

In just the last century, hundreds of tonnes of man-made structures started hurtling thousands, and then, relatively quickly, millions of people across the world through the skies. Initially it was not safe. Only the brave would dare fly. However, once Boeing developed the 747 Jumbo Jet, the concept of flight for every human being became a possibility. Now it takes bravery on the part of the authorities to stop people from flying.

The development of the Boeing 747 was the turning point for the aviation MRO industry. The process of maintenance was formally defined for the first time, thus making flying safe for the general public. The advent of standardised aircraft maintenance started with the formation of the Maintenance Steering Group (MSG). The history and details of this group is publicly available, therefore we will not delve into further details of its constituents (i.e. the OEMs, FAA and others) and its mission.

However, it will be essential to revisit the background of MSG-3, which is currently enforced on the aviation MRO industry; in the Air Transport Association's (ATA) words:

> In July, 1968, representatives of various airlines developed Handbook MSG-1, 'Maintenance Evaluation and Program Development,' which included decision

logic and inter-airline/manufacturer procedures for developing scheduled maintenance for the new Boeing 747 aircraft.

Subsequently, it was decided that experience gained on this project should be applied to update the decision logic, and to delete certain 747 specific detailed procedural information, so that a universal document could be made applicable for later new type aircraft. This was done and resulted in the document, entitled, 'Airline/Manufacturer Maintenance Program Planning Document,' MSG-2. MSG-2 decision logic was used to develop scheduled maintenance for the aircraft of the 1970s.

. . . ATA airlines decided that a revision to existing MSG-2 procedures was both timely and appropriate. The active participation and combined efforts of the FAA, CAA/UK, AEA, US and European aircraft and engine manufacturers, US and foreign airlines, and the US Navy generated the document, MSG-3.[6]

The universal aircraft maintenance process is based on MSG-3 and I would recommend that all serious students of aircraft maintenance and other interested readers make themselves familiar with the document, published by ATA.

In the following paragraphs we discuss the aircraft life cycle and then the aircraft maintenance life cycle for each domain. Aircraft maintenance is part of the aircraft life cycle and is discussed in detail. Figure 3.1 shows the aircraft life cycle from its maintenance perspective.

From an aircraft maintenance perspective there are four domains, which have their own life cycles and which live within the overall aircraft life cycle:

- airframe and structural

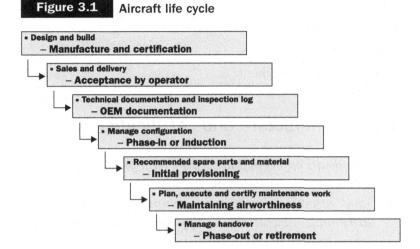

Figure 3.1 Aircraft life cycle

- engine

- component

- GSE (ground support equipment).

All four domains come together to support the maintenance of an aircraft. Their life cycles might not follow each other strictly but they are intertwined such that one cannot survive without the others.

Many activities take place before an MRO organisation gets its hands on the aircraft. However, we will limit our discussions to those that are relevant to maintenance and elaborate these within the aircraft life cycle. These activities are:

- manufacture and certification

- acceptance by the operator

- OEM documentation

- phase-in or induction
- initial provisioning
- maintain airworthiness
- phase-out or retirement.

3.2.1 Manufacture and certification

Certification: Formal procedure by which an accredited or authorized person or agency assesses and verifies (and attests in writing by issuing a certificate) the attributes, characteristics, quality, qualification, or status of individuals or organizations, goods or services, procedures or processes, or events or situations, in accordance with established requirements or standards.[7]

Manufacturers design aircraft, engines and components with explicit maintainability in mind. They achieve a high degree of maintainability by adhering to MSG-3 logic and complying with the relevant regulatory instructions and standards. The operators and the buyers, which are typically airlines or fleet operators, also participate in the design process and provide maintenance related feedback and concerns based on their experience.

The aircraft require 'type certification' before they can be flown. This is the beginning of the airworthiness of an aircraft type. The certification tells the world that this type of aircraft is safe to fly in. An aircraft which is type certified can be sold to the operators, who can operate it to carry passengers. The buyers are the airlines/operators or the leasing agencies. The buyers, once they take delivery of the aircraft then take on the onus of the legal obligation to keep the aircraft airworthy as specified in the type certification.

The Type Certificate Data Sheets (TCDS) database is a repository of Make and Model information. The TCDS is a formal description of the aircraft, engine or propeller. It lists limitations and information required for type certification including airspeed limits, weight limits, thrust limitations, etc. TCDS documents are available from the initial implementation of this database (April 1999) forward. Historical documents prior to implementation are not necessarily included in this database; however, historical information will be retained as additional documents and revisions to documents are incorporated into the database. Revision History refers to those documents that have been revised through a TCDS revision. Current documents refer to those that are the most current. When looking at the TCDS and Model information in any of the Revision History views, the checkmark indicates the current documents.[8]

According to the FAA: 'An airworthiness certificate is an FAA document which grants authorization to operate an aircraft in flight.' The details of the certification are available on the FAA website. The specific document for type certification is FAA Order 8130.2F.

3.2.2 Acceptance by the operator

Acceptance: Buyer's approval of the goods supplied at their invoiced price, signified by the act of taking delivery, and/or use, of the goods without protest.[9]

An operator can accept an aircraft directly from the manufacturer, or from a leasing agency, or another operator. Acceptance by an operator implies that either the operator themselves or a competent agency, which can be an MRO organisation, has inspected the aircraft and the aircraft is airworthy.

An aircraft acceptance process includes but is not limited to: review of all aircraft documentation, modification summaries, service bulletins and airworthiness directives, etc. One or more acceptance flights are conducted, with the manufacturer's representative on board, to ensure that all aircraft systems operate correctly and the aircraft meets all airframe specifications and engine performance parameters. The acceptance process may even include a detailed cosmetic inspection of the exterior paint, interior fabrics and woodwork, and an operational check of the cabin entertainment systems and any other installed systems. Normally a team of experts is sent to the manufacturer's site to evaluate the aircraft and once the aircraft is accepted the team flies back with the aircraft, ready to put it into service.

3.2.3 OEM documentation

In engineering, technical documentation refers to any type of documentation that describes handling, functionality and architecture of a technical product or a product under development or use. The intended recipient for product technical documentation is both the (proficient) end user as well as the administrator/ service or maintenance technician. In contrast to a mere 'cookbook' manual, technical documentation aims at providing enough information for a user to understand inner and outer dependencies of the product at hand.[10]

An operator receives a significant amount of technical data from the OEMs in electronic form. These come from aircraft, engine and component manufacturers, and the sheer volume of technical information provided must not be underestimated. It is said that once, in the 1980s when a fighter plane was delivered to the US Department of Defense, it came with around 18 tonnes of documentation, which inspired the DoD to contribute to funding the development of SGML, the father of HTML and XML, so that the data could be transmitted electronically and would not require a truck to transport it.

ATA and Boeing developed a standard called ATA 2100, which was based on SGML, to standardise all the OEM documentation. This standard is now merged with ATA's iSPEC2200 and is expected to evolve into S1000D. These standards have helped reduce the complexity and volume of the data transmission. However, they require proprietary software to receive, store and manage the documents. It is hoped that in time the aviation industry will adopt the XML standard fully, which will allow the use of generally available software.

The good thing about the structure of the maintenance-related data is that it uses a well defined convention to specify the tasks that need to be performed for maintenance. This convention is based on AMTOSS (Aircraft Maintenance and Task Oriented Support System). As mentioned above, the overall standards used for documentation are iSPEC2200 and S1000D. The discussion on these standards is outside the scope of this book. The technical documentation received is (the list is not exhaustive):

- IPC (Illustrated Parts Catalogue) configuration data
- MPD (Maintenance Planning Data)
- AMM (Aircraft Maintenance Manual)
- EM (Engine Manual)
- CMM (Component Maintenance Manual)

- Time Limits Manual
- Tools and Equipment Manual
- FIM/FRM (Fault Investigation Manual/Fault Reporting Manual).

One of the many documents that the operator receives is the Inspection Log, which describes all the MSI (Maintenance Significant Items) fitted to the aircraft. This in effect contains all the relevant information on the aircraft configuration. The seller is obliged to provide this information to the buyer.

Currently, all the relevant aircraft-related documents are IT enabled. This relieves some of the legal pressure that the OEMs have for distributing hard copies of the manuals to all the operators, and keeps them current. The OEMs actually give incentives to the operators for adapting to the electronic format. These incentives include free software and sometimes access to their servers.

3.2.4 Phase-in or induction

> Induction: Implementation strategy in which new equipment, policies, or processes are introduced . . .[11]

When an aircraft arrives and before it is placed into service for the first time, a lot of information gathering, collating and storing of data is accomplished by the people involved in its maintenance. This is the first phase of the aircraft maintenance lifecycle as well and is called 'Phase-in' or 'Induction' of the aircraft. This ritual of identifying the aircraft is the beginning of the configuration management of an aircraft.

Configuration management is one of the most challenging jobs for an operator and the MRO organisation assigned to do this exercise. The purpose of this exercise is to keep and manage the records of all the effects of maintenance on the aircraft and its components.

Essentially the responsibility of configuration management lies with the operator, typically an airline. This is because legally the operator of an aircraft is responsible for its airworthiness, hence the configuration management. However, traditionally the assigned MRO organisation undertakes this responsibility.

At this stage an MRO organisation is normally aware of the aircraft type that is being inducted. If not then the effort required to configure an aircraft becomes substantial. In the case where the organisation has previous experience of maintaining this type of aircraft it draws upon its knowledge of the maintenance concepts applied to the aircraft type. Otherwise, it has to identify the OEM's applicable maintenance concepts and create a set of options for the maintenance program. This is done using the information provided by the OEMs.

As discussed in the preceding section, the OEMs provide these documents: IPC configuration data; MPD, AMM, EM, CM task data, FIM/FRM data, and SRM tasks. The documents are electronically received and processed and are evaluated for long-term fleet maintenance and engineering requirements, which help in long-term forecasts for resources. Based on the evaluation there may be a need to modify/tailor these documents to meet specific operator's requirements. (The management of the technical documentation and its publication is described elsewhere in this book.)

The information received helps an operator or the MRO organisation create a virtual model of the aircraft type within its information system. The process of building the model is very complex, if started from scratch. Various constraints need to be catered for to build a perfect model. The part-numbering scheme, used in aviation industry, is based on iSPEC2200 standards, but most of the currently available ERP systems do not cater for this specific standard. In SAP for example a special module has to be implemented to cater for this requirement. There are requirements for multi-level interchangeability of the parts, e.g. hierarchical interchangeability and multiple alternatives for complex ownership of the parts, etc.

An operator or the MRO organisation loads and manages the information from the OEMs independently. The configuration-related information alone is so large and complex that some of the niche MRO software cater to just this requirement!

The configuration and the preparation of an aircraft for induction go through the following steps:

- The data from OEMs is loaded as draft data; however, it must be recognised that this data is not static and undergoes change based on the actual configuration of the aircraft.

- A reference state, i.e. point in time, is established for the configuration. This is because the aircraft does not sit idle while it is being configured in the system. The configuration work starts, or should start, before the aircraft has left the manufacturer or the previous owner's site.

- While configuring the aircraft, it is necessary to ensure that component part numbers are unique, especially within sub groupings such as a specific manufacturer, and all the tracking related information (e.g. airframe serial, tail number, line number, customer number, block number, etc.) are recorded. The part numbering scheme, as mentioned earlier, follows iSPEC2200 and in the case of the 787 it is S1000D. Additionally, the part characteristics (dimensions, handling information, hazard and risk assessment information, material data, safety sheets, etc.) are recorded.

Once the configuration of an aircraft type is recorded then the actual aircraft is modelled in the system. This is done by simulating the fitment of actual serial numbers to the virtual model created earlier. The information about which serial numbers are actually fitted in the real aircraft comes from the manufacturer's Inspection Log, when the aircraft is inducted for the first time. Later on, it is the responsibility of the operator to pass on the latest information on what is fitted on the aircraft to the MRO organisation before maintenance planning can begin. This exercise is also known as induction or phasing-in of an aircraft, engine and components. We discuss these concepts in more detail later in this chapter. Meanwhile, we should be aware that the input and the methods of managing the configuration for each of them are different.

These activities establish a base line or draft data to model the aircraft in the system. A team of engineers/technicians are then required to refine the draft data. The activities may include:

- Set up of an allowable configuration hierarchy.
- Identify the functions of the components.
- Set up specific part numbers for each function, specifying quantity and preferred part number, etc., to be operational as per MEL/CDL.
- Set up the actual serial numbers fitted in the aircraft.
- Record the modification status of the aircraft, engine and components.
- Record the history of changes (the information currently comes mainly from the Inspection Log).

This way a full configuration model for an assembly (e.g. aircraft) from sub-assembly configurations (e.g. engine) is created. Further, fitment rules are established, which specify and enforce how the parts can be fitted in a mixed mode and/

or according to cross functionality. There are many other minor activities which are not covered in the steps above. However, they are mostly subsets of the activities described above, and therefore are not discussed in this book.

Once an aircraft is configured in the system it lives there until the aircraft, engine or the component is phased-out. Normally the MRO organisations and, in most cases, the operators have their own systems which are sophisticated enough to capture and process the information generated by phase-in/phase-out processes. However, the leasing companies do not have any compelling reasons to own such a system. They mostly rely on the MRO organisation for the relevant information.

The final step in the configuration management is to identify and report any deferred and anticipated defects. These defects do not necessarily affect airworthiness of the aircraft.

We notice here that this set of activities is simply about information gathering and recording, and in some cases these are repetitive and valid across operators. In order to maintain the integrity of an aircraft it is essential to record and maintain accurate configuration.

3.2.5 Initial provisioning

According to the (US) DOD, Initial Provisioning: The process of determining the range and quantity of items (i.e., spares and repair parts, special tools, test equipment, and support equipment) required to support and maintain an item for an initial period of service. Its phases include the identification of items of supply, the establishment of data for catalogue, technical manual, and allowance list preparation, and the preparation of instructions to assure delivery of necessary support items with related end articles.[12]

In order to maintain the airworthiness as designed in the aircraft, the operator is obliged to keep a minimum amount of inventory of spare parts, which the manufacturers recommend. An operator does not have to abide by all the recommendation from the manufacturer. In fact it is the MRO organisation that ends up procuring and storing the spare parts. In order to have the right levels of inventory, the MRO organisation has to plan and procure an agreed quantity of spares. However, this agreement is between the operator and the local Civil Aviation Authority. There are a whole set of techniques used in initial provisioning and we will discuss the techniques related to procurement and materials management of spare parts later in this chapter. The spares parts, initially provisioned, are mandatory and they are linked with the airworthiness of the aircraft.

The recommendation from the OEMs specifies the items and the suggested quantity. The information also contains the other inventory management-related information, which helps the operators to strategise their procurement.

The main task of the Material Management group, involved in the initial provisioning, is to first identify the spares that they might already have in their stores and map the recommended quantity to the current inventory management parameters. This ensures that there are no duplications. After arriving at the list of existing parts, the Procurement Department identifies the new parts and subsequently the vendors supplying these parts. Some of the new parts may be supplied by the existing vendors. These parts are then recorded as the new items for these vendors. Finally, the parts and vendors might both be new, which will require going through the vendor negotiation and contracting process.

The most important point to note is that there is an extra parameter added in the initial provisioning exercise for aircraft: the minimum quantity from the point of view of airworthiness. The Procurement Department of an MRO organisation does not have much say in this. Otherwise the procurement process for aircraft spares is similar to those used in other industries.

In addition to the above, the OEMs, based on their experience, try to predict replacement frequencies and time to failure. However, these values are not necessarily the best suited to an operator. Therefore the operator and the MRO organisations conduct their own analysis and arrive at the quantities of the spare parts that will be required for replacement, when and from whom.

3.2.6 *Maintaining airworthiness*

Maintenance: Actions necessary for retaining or restoring an equipment, machine, or system to the specified operable condition to achieve its maximum useful life. It includes corrective maintenance and preventive maintenance.[13]

An aircraft is designed to fly – it is airworthy and carries a certificate for it. However, as it travels the skies its structure and components deteriorate. This deterioration leads to a condition which renders the aircraft un-airworthy. This is where the process of maintenance kicks in. The act of maintenance is performed to bring the aircraft back to a condition where it can be allowed to fly, i.e. render it airworthy again.

Since an aircraft cannot afford to break before it is fixed, it needs to be inspected or checked regularly and frequently. These checks are so pervasive that all the maintenance activities are grouped and named as 'Checks'. An aircraft is checked before and after each flight, which can sometimes be as simple as a visual inspection and a walk-around. It is also checked during a night halt and at various intervals predetermined by the maintenance plan created for the aircraft.

The cycle of maintenance, described in detail later in this chapter, starts with inspection, then describes rectification and finally certification. The types of maintenance carried out are: Line or Ramp maintenance and Heavy or Hangar maintenance. The maintenance is planned such that the aircraft should not be called in for repair because of any unexpected defects. The strategy of maintenance was discussed in Chapter 1.

The aircraft maintenance life cycle for ensuring the aircraft's airworthiness is made up of three independent life cycles: (1) Airframe Maintenance Life Cycle, (2) Engine Maintenance Life Cycle and (3) Component Maintenance Life Cycle. There is, however, one more life cycle, which we shall discuss later in this chapter: Ground Support Equipment Maintenance Life Cycle. The reason I have separated this life cycle from the other three is that the GSE life cycle is not directly attributable to an aircraft's airworthiness.

It should be remembered that the engine and component maintenance life cycles are also identified as 'off-wing' maintenance life cycles whereas the airframe maintenance life cycle is called 'on-wing'. The treatment of these life cycles later in this chapter will elucidate further the difference between these two types of maintenance life cycles.

3.2.7 *Phase-out or retirement*

Retirement: Removal of an asset, equipment, property, or resource from service after its useful life, or following its sale.[14]

The reason for the phasing out of a component or the aircraft could be various: through obsolescence, or when sold to another operator, scrapped for regulatory compliance, unserviceable. There is a specific document that the MRO/ operator organisation is obliged to create and deliver for archiving or send to the buying operator, who is the new owner. In earlier days, the sale of an aircraft in the secondary market used to be a transaction between the two involved operators. The selling operator phases out the asset whereas the buying operator phases in the asset. However, now that the leasing companies have come in vogue and happen to be interested in the sale transaction, they are also interested in the phase-in and phase-out processes.

In general an aircraft is phased out for two reasons:

- The operator sells the aircraft to another operator or a leasing company.
- The operator decides not renew the lease of a leased aircraft.

In both conditions the operator is obliged to conduct a final check and provide configuration information to the receiving party.

An aircraft is retired once the Regulatory Authority declares it unfit to operate either due to age or maintainability. In either case the operator has to follow a closure procedure and document the status for record. The aircraft is then cannibalised and mothballed.

In case the aircraft is destined to be operated by another operator, it has to be thoroughly inspected. The operator then transfers the configuration-related information to the buyer or the leasing company. One can take it as if the operator creates the information akin to an OEM because the next step for the new operator is to phase in or induct the aircraft.

In other words, the information flow out of the phase-out phase should enable the buyer to conduct the phase-in phase.

As we discussed earlier, at the end of an aircraft's life, either due to sale, non-renewal of lease or retirement, a final check is performed to record its current configuration and defects.

In case of sale or non-renewal of lease, the aircraft goes through the maintenance process to ensure that it is airworthy. All the necessary information related to the aircraft is sent to the receiving party either electronically or on paper. Normal practice in the industry is to make this activity coincide with a major check, normally a C-check.

In case of retirement the serviceable components are identified for cannibalisation and a specified mothballing exercise is conducted. A mandatory set of information is also sent to the regulatory authorities for archiving.

Finally, the operator de-registers the aircraft and conforms to the processes required by the regulatory authorities and hands over the aircraft to the party concerned.

3.3 Airframe maintenance life cycle

Figure 3.2 shows the life cycle of airframe maintenance. This life cycle is valid for any MRO organisation, in-house or external. It must be noted that the life cycle does not include long-term demand and resource planning for the maintenance of airframes, engines and components. We assume that at this stage the MRO organisation has completed the exercise

Figure 3.2 Airframe maintenance life cycle: aircraft visit

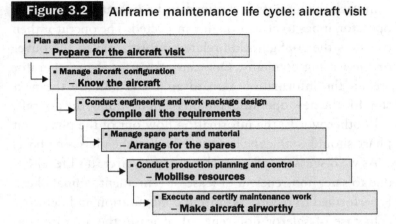

of long-term planning, which we covered in the previous chapter. Moreover, we further assume that the MRO organisation we are referring to is a full service organisation, which may not be the reality in many cases.

In this section we discuss airframe and structural maintenance. We will discuss engine, component and GSE maintenance life cycles in the later sections.

It is, however, worthwhile reiterating that the maintenance of airframes and structures falls under the most rigorous regime of compliance. This is the most important segment of the maintenance process. Almost all the items maintained thus fall under the 'on-wing' category. The cycle and interplay of the functions remain almost the same as described above, where the aircraft maintenance starts with scheduled or unscheduled tasks and ends with certifying the work.

The functions or phases of the Airframe Maintenance Life Cycles are:

■ Prepare for the aircraft visit: plan and schedule work.

■ Know the aircraft: manage aircraft configuration.

■ Compile all the maintenance-related requirements: conduct engineering and work package design.

- Arrange for the spares: manage spare parts and material.

- Mobilise resources: conduct production planning and control.

- Make aircraft airworthy: execute and certify maintenance work.

The functions identified above are discrete and more or less independent of each other, in the sense that these functions can also be defined as independent services. Hence, one can automate all the functions in each phase of the life cycle separately, by deploying separate and dedicated modules of application software. In fact, it can be architected keeping interoperability in mind such that it supports all the functions of the phases. Some software vendors have attempted just that. They have created software to cater to just one phase with the ability to integrate with other MRO applications.

Before we dive into the world of maintaining the aircraft, we should keep one simple dictum in mind. Once a component or part is removed from a serviceable or airworthy aircraft, the component is rendered or declared unserviceable. In order for the removed part to be fitted to an aircraft again, it has to go through the ritual or cycle of being made serviceable. This is irrespective of whether the part or component was faulty or not. The purpose of this regime is to ensure that an unserviceable part is never fitted on to an aircraft.

3.3.1 Prepare for the aircraft visit

Actions: Conduct short-term capacity and resource planning, allocate resources and schedule jobs.

Though this phase of aircraft maintenance is generic, the treatment for aircraft, engines, components and GSEs are different from each other. Hence, we will discuss this phase separately for each type in the following sections. The common activities in this phase are identification and validation of requirements, technical as well operator specific; identification of capacity and capability vis-à-vis requirements; and deliberate short to mid-term maintenance planning.

The major activities in this phase are:

- Aircraft maintenance scheduling.
- Line/Ramp maintenance resource planning.
- Hangar and slot resource planning.

The objective of this phase is to ensure that the MRO organisation is not just willing but also prepared to receive the aircraft for maintenance. This is a very critical phase for the organisation as these mid- and short-term plans have tremendous impact on the profitability of the organisation.

The MRO organisation must have the aircraft's routing information, i.e. when and from where the aircraft will arrive. This allows the MRO organisation to plan when it should take up the maintenance operation and where: in the hangar or the ramp.

The MRO organisation expects to receive the following information regarding the aircraft:

- Routing: when it is scheduled to fly into the location.
- Maintenance requirements: both from OEM's perspective and the operator's.
- Maintenance work packages: if prepared by the operator.
- Deferred defects: the non-critical snags that need to be fixed during the schedules maintenance.
- Mandatory modifications: all applicable service bulletins and ADs issued by the OEM and accepted by the operator.

The MRO organisation plans and schedules hangars/slots/ ramps according to the scheduled/unscheduled arrival of the aircraft; and records and manages service offerings and resource pool, including skilled personnel, tools and supporting items.

The performance of the MRO organisation's planning capabilities is measured by the following metrics:

- Accuracy of the plan: measured by the deviations that the original plan had to go through during production planning and control.

- Speed and timeliness of plan: how long it takes to generate an aircraft visit plan.

3.3.2 Know the aircraft

Actions: Manage aircraft configuration; identify record and update aircraft's configuration-related information.

From the first time that the aircraft had been configured in a system, the plane normally undergoes a lot of changes. It is akin to our body losing cells to reinvigorate it. A significant number of parts are removed and new parts fitted and many different materials are replaced as the aircraft flies from airport to airport. These changes are captured either electronically or manually. It is mandatory to capture them, hence the MRO organisation requires the latest configuration-related information of the aircraft to effectively carry out maintenance.

Legally, it is the operator of the aircraft who is responsible for keeping the configuration data up to date. However, it is the MRO organisation which needs the correct data for ensuring the airworthiness of the aircraft.

In the case where the MRO organisation is part of the operator's organisation, then this data is perennially available for maintenance. But if the operator has outsourced the maintenance then either the operator outsources the system as well or supplies the data to the MRO organisation prior to the aircraft visit. This creates an interesting conundrum. Should the operator maintain a system to manage the configuration of the aircraft or depend on an MRO organisation's system? Currently, there are many different types of information exchange that take place for the configuration data between the two organisations. The operators who depend on one large MRO organisation for their major maintenance needs tend to relinquish the responsibility of maintaining the configuration data. This strategy, in some cases, limits their ability to flexibly change the maintenance vendors. The operators become 'tightly coupled' with a specific MRO organisation. There are benefits in this approach, which in some cases far outweighs the loss of flexibility.

In any case, the MRO organisation requires the aircraft's configuration information, which only the operator can provide. This information is required as a starting point. In other words the MRO organisation identifies the aircraft and records its details to conduct the step described in the next section.

3.3.3   Compile all the requirements and create a maintenance plan

This process is the core of maintenance execution. It includes modification and defect rectification requirements.

The aircraft operator specifies the requirements and the MRO organisation does an impact analysis to arrive at baseline maintenance requirements, which can be scheduled, executed and monitored after an agreement between the two organisations has been concluded.

Once the requirements are firmed up the first task is to identify maintenance opportunities. In other words, mapping of the requirements with the organisation's service offerings and evaluation of all the requirements whether they can be met or they need support from partners or third party vendors. This exercise of requirement mapping happens for every planned visit of an aircraft, first time or not.

Normally, an aircraft visit is for major scheduled maintenance work. In older terminology, it is generally for heavy maintenance. We will discuss both the line and heavy maintenance and how the Maintenance Control Centre (MCC) orchestrates the activities.

An aircraft comes with four different types of maintenance requirements:

- *Scheduled maintenance program*: This is provided by the operator and is part of the Maintenance Plan created when the aircraft was inducted.

- *Modification requirements* (MODs): They are of two different types but originate as Service Bulletins (SB). They are either mandatory or advisory. The operators decide, sometimes with the support of the MRO organisation, which MODs are to be accomplished on this visit.

- *Deferred defects*: These are the defects which the operator had identified but decided to rectify later and has been packaged for this visit. Generally, these defects are not related to airworthiness.

- *Unscheduled/non-routine tasks*: In the real world we know that it is impossible to predict all the maintenance requirements beforehand and a lot of defects are found once the aircraft panels are opened or an inspection is conducted. These can be as much as twice the number of defects that have been planned for rectification.

All the above-mentioned requirements, barring the last one, are compiled and reviewed, and a contingency is set up for the unscheduled requirements. These requirements are always in terms of Tasks and specified in the Task Cards, which are standard across the industry. Most of them are originally created by the OEMs, however some of them are modified by the operators. The MRO organisation normally does not alter or create Task Cards. These tasks are then grouped as work packages and estimated for the resource requirement. The MRO organisation deliberates the cost of each work package and gets approval from the operator.

The approved work packages are validated by engineers and sequenced such that they fit the workflow and panel opening and closing. Resources are allocated and levelled for the work packages and a maintenance schedule is created.

Each Task Card contains the effort, material, and tool requirements. These are mainly recommendations from the OEM but the MRO organisation is expected to keep its own historical data to apply them to the task. This is an area of contention between an operator and the MRO organisation thus may have different assumptions and values for the same task. However, these are resolved during the contract negotiation.

3.3.4 Arrange for the spares

Actions: Plan and manage procurement, inventory and logistics of spare parts and material.

Let us talk about managing spare parts and materials which are an integral part of maintenance activity. It seems like an obvious statement but when it comes to managing these spares, especially for maintaining an aircraft, it turns out that it is not so obvious. Why do I say that?

Because these spares are not any ordinary spares. They have to be serviceable and they should contribute towards the airworthiness of the aircraft to which these are to be fitted.

Let us work on this. An aircraft visit is scheduled; the Engineering Planning Department has prepared the work package; identified and sequenced the Task Cards; captured the material requirement for each task from the Task Cards; and raised the material requisitions. As far are the engineering planning is concerned, their job is over.

Now in comes the Material Management Department. Here they are no different from any other supply chain organisation. Their goal is to fulfil the material requisitions. However, the objective of the Materials Management Department remains provisioning of the spare parts for aircraft maintenance at the right time and at the right place.

There are various strategies that the Materials Management Department uses, e.g., preloading of spares for scheduled work, anticipating unscheduled requirements, quick response to ad hoc requirements by reducing lead-time for procurement, etc. In the following paragraphs, we touch upon the activities required to meet the above stated goal of the Material Management Department. There are many books and papers on the subject, so here we will discuss the aviation MRO point of view only.

3.3.5 Mobilise resources

Actions: Allocate hangar slots, tooling, support equipment, technical staff; sequence and schedule Task Cards for maintenance execution.

The production schedule, as I prefer to call it as 'maintenance schedule' implies non-sequential operations, is then broken down into set of tasks based on the types of resources required, e.g. mechanical, airframe, electrical, etc. These groups of tasks are distributed to the workforce.

Similar to a resource planning exercise in any industry, matching the requirement and the availability of resources for an MRO organisation is always a challenge. Due to the stringent requirement of the quality of skills all the resources that are deployed must be certified and in many cases licensed. In the areas of maintenance where airworthiness is concerned, it is illegal to deploy someone who is not licensed. The licensing process is a long drawn out affair and an MRO organisation has to ensure that the right number of licensed engineers and technicians are available for the aircraft visit.

The specific skill requirements are mentioned in the Task Cards, which are used for resource planning and resource levelling.

3.3.6 Make aircraft airworthy

Actions: Distribute work, record actual hours and material consumption, support trouble shooting, rescheduling due to non-routine requirements and get the final certification of airworthiness.

A technician opens the Task Card, studies it, gathers the material and the tools specified, and records the starting time. Once the task is complete, he records the ending time and signs off the task Card. There may be situations where the task may not be complete at the end of a shift. In this case the technician closes the Task Card and hands it over to his supervisor. These Task Cards are then reopened by the supervisor of the next shift who reallocates them to the new crew.

The production planning and control team keeps track of the task execution and its progress.

In many cases there are no options but to resort to overtime. This creates an interesting situation, because it not only increases the cost of maintenance but also limits the availability of some highly skilled resources. This is an area which requires detailed discussion and IT can be leveraged to reduce overtime and consequently reduce the cost of maintenance. This 'overtime' is the biggest headache of an MRO organisation. It literally bleeds money. It is believed that if the organisation can control the overtime, it does not have to bother about anything else to become highly profitable. This, however, is easier said than done.

No MRO organisation that I am aware of tries to measure and correlate the overtime spend with its performance, which in my opinion should be done. The information is available and a dashboard can be set up as an alert mechanism and then a follow through of the causes will help the organisation stop this bleeding.

Obviously, the main thrust of the organisation at this juncture is to planning and scheduling. A robust and optimised schedule at the beginning of maintenance work and then the ability to rapidly re-optimise the schedule will ensure the efficiency of delivery, i.e. making the aircraft airworthy.

The process works like this. The Task Cards, in a work package, are assigned to a specific technician, who records the start time, performs the task as specified, records the end time and either signs off the task himself or by an authorised person. The life cycle of a Task Card is: open task, perform task and close task. Any defect or discrepancy found while the task is being performed is reported and another Task Card is assigned to rectify the defect. No task can be performed without a Task Card.

The end game of this task oriented work is that all the tasks are signed off. These signed-off Task Cards become the basis for the final airworthiness certification of the aircraft. Once the aircraft is certified, it is then available for operation and its tail number can be assigned to the flight number/s that it will operate, before coming back for maintenance.

3.4 Aircraft engine maintenance life cycle

Figure 3.3 shows the life cycle of engine maintenance, which is valid for any MRO organisation, in-house or external. This does not include long-term demand and resource planning for the maintenance of engines. We assume that at this stage the MRO organisation has completed the exercise of long-term planning, which we discussed in the previous chapter. Moreover, we further assume that the MRO organisation we are referring to is a full service organisation and conducts engine maintenance, which may not be the reality in many cases.

In this section we discuss aircraft engine maintenance. We will discuss component and GSE maintenance life cycles

Figure 3.3 Engine maintenance life cycle: engine overhaul/ repair

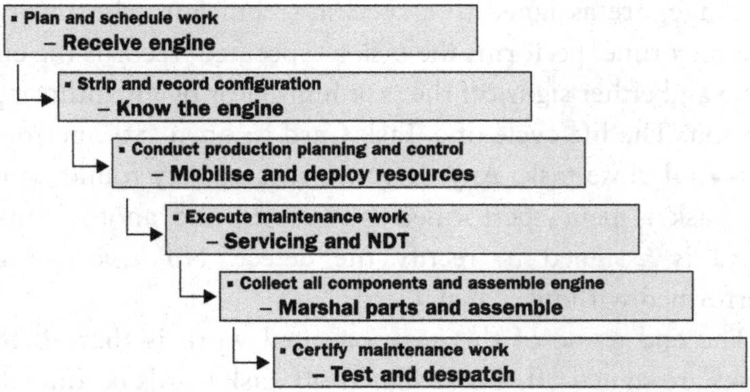

- Plan and schedule work
  - **Receive engine**

- Strip and record configuration
  - **Know the engine**

- Conduct production planning and control
  - **Mobilise and deploy resources**

- Execute maintenance work
  - **Servicing and NDT**

- Collect all components and assemble engine
  - **Marshal parts and assemble**

- Certify maintenance work
  - **Test and despatch**

in the subsequent sections. It is, however, worthwhile reiterating that the maintenance of aircraft engines does not fall under the rigorous regime of airworthiness compliance. Almost all the items maintained thus fall under the 'off-wing' category. The cycle and interplay of the functions remain almost same as the aircraft maintenance life cycle, where the engine maintenance starts with scheduled or unscheduled tasks and ends with certifying the work.

The functions or phases of the aircraft engine maintenance life cycle are:

- Receive engine: plan and schedule work.

- Know the engine: manage aircraft configuration.

- Mobilise and deploy resources: conduct production planning and control.

- Servicing and non-destructive testing (NDT): execute maintenance work.

- Marshal parts and assemble: collect all components and assemble engine.

- Test and despatch: certify maintenance work.

The functions identified as phases, above, are discrete and more or less independent of each other, in the sense that these functions can also be defined as independent services. Hence, one can automate all the functions in each phase of the life cycle separately, by deploying separate and dedicated modules of application software. In fact, it can be architected keeping interoperability in mind such that it supports all the functions of the phases. Some software vendors have attempted just that. They have created software to cater to just one phase with the ability to integrate with other MRO applications.

Fortunately for engine maintenance crew, the removal of a part or component from an engine does not render the part or the component unserviceable as it does for the aircraft. Hence, unless until it fails a test or has a defect reported, a removed part or component from an engine in the engine shop is considered serviceable. However, here we assume that the component is being removed from an 'off-wing engine'.

The issue of rendering a component or a part serviceable is important because it makes recording of the engine stripping information much simpler. In other words, this activity is no different from any other plant maintenance process where disassembly and assembly of components are involved.

However, engine maintenance is similar to aircraft maintenance on the following counts:

- The maintenance plan is MSG-3 compliant.
- The maintenance activity is task oriented and uses AMTOSS Task Cards.
- The engine requires an airworthiness certificate before being despatched after maintenance.
- Engine maintenance requires extensive record keeping and preservation of historical information.
- The maintenance is strictly carried out as specified in the OEM manuals (EMM, FIM, etc.).

3.4.1 Receive engine

Actions: Plan and schedule work; conduct short-term capacity and resource planning; allocate resources and schedule jobs.

Once an engine arrives at the engine repair and overhaul shop, it is removed from the carriage, which is specially designed for engine transportation, and brought into the receiving bay. Here the related paperwork is done: matching the engine serial numbers and ensuring that the information related to its current status are available.

The details of the engine are recorded and updated. In case this is the first visit of the specific engine the paperwork involved will take longer than for an engine known to the shop. Yes, the engine shop does receive the same engine again and again.

At this point the engine shop triggers off the production planning exercise based on its knowledge of the engine type. The work packages, consisting of AMTOSS Task Cards, are created to execute the maintenance activities.

The engine bays are generally moveable structures with easy accessibility for disassembly. In most cases you will see the documentation of the stripped engine stuck on the side of these structures.

The information required at this stage is generally available with the operator or the MRO organisation, which has sent the engine for repair/overhaul. In some cases the engine shop itself may have the basic information on the engine concerned.

3.4.2 Know the engine

Actions: Strip and record configuration; identify record and update engine's configuration related information.

The 'received' engine may be known to the organisation or may not be. If it is a first time visit there might not be any previous records available. However, it is a general practice of an engine shop to recreate the configuration of an engine by actually stripping it down and then recording it. Nowadays the engines are very modular. Hence stripping down is a regular and repeatable operation.

Recording of every stripped component and part from the engine is required to ensure that correct items are repaired and then re-assembled.

Here let us note an interesting item. When an engine is 'on-wing' it becomes part of the aircraft configuration, which an engine shop need not be aware of. The affect of this is that when any component is removed from an 'on-wing' engine, it becomes 'unserviceable' and needs to go through airworthiness certification before becoming 'serviceable'. In practical terms this does not make much difference, but from the point of view of storing information it does. The regular dilemma faced by an information system is: should an aircraft configuration system keep in-depth engine configuration or not. In most cases it does not. And it so happens that engine shops also do not keep in-depth configuration information. However, it is the OEMs who do.

3.4.3 Mobilise resources

Actions: Conduct production planning and control; plan and manage procurement, inventory and logistics of spare parts and material.

An engine shop has standard work breakdown structures for the activities that are normally carried out for servicing or overhauling an engine. Nonetheless, in order to arrive at a workable and optimised schedule the engine configuration information, gleaned from stripping the engine, is required. In addition to the configuration data, further information such as identified or deferred defects information is taken into account.

The production planning and scheduling department then creates a schedule of activities based on the constraints, such as type of service, defect rectification and special modifications to be carried out.

The production schedule thus created becomes the driver for resource allocation. The production plan then calls for the material required, the type of skills to be deployed and for how long.

The process adapted for production planning and control for engines is similar to the aircraft, where work packages are created by grouping routine, non-routine and modifications-related Task Cards.

3.4.4 Service and non-destructive test (NDT)

Actions: Execute maintenance work; clean and conduct non-destructive testing on the aggregates and other components of the stripped engine.

All engine shops are equipped with standard NDT facilities and tools. The EMM (Engine Maintenance Manual) specifies which components of the stripped engine need testing. There are copious amounts of information available on the actual process for the readers to refer to. However, here we look at this process from the point of view of information collection and application.

There are two sources of information that help identify which components are to be tested and when. The MPD specifies 'when' based on the hours or 'cycle' run by the engine. The EMM specifies how the testing is to be done and what the criteria are for pass or fail. The testing activities are factored into the production plan and scheduled for execution.

The components that fail the test are taken out and disposed of and a request for the replacement component is placed.

The main component of the information collected at this stage is pass/fail, binary information. However, the test results are also captured for later 'reliability' analysis, which affect maintenance related parameters, e.g. mean time between failures.

3.4.5 Marshal parts and assemble

Actions: Collect components and assemble engine; bring all the cleaned and tested parts of the engine together, procure replacement parts and assemble the engine.

To marshal literally means: 'put into a proper or systematic order'. In the context of engine overhaul or servicing, I redefine the activity as: put the serviceable parts into a proper and systematic order.

During the actual execution of the servicing activities carried out as specified in the Task Cards, one would notice that the components are cleaned and inspected for wear and tear. Not much repair work is done. If there is a need for repair then normally the component is sent out to the component repair shop or to a third-party repairer.

The serviceable components are 'marshalled' from the stores and the servicing area for assembly. Once the full configuration criteria are met the engine is re-assembled.

From an information point of view this process requires very detailed and accurate information. Lack of correct information often results in major delays in re-assembly of the engines.

3.4.6 Test and despatch

Actions: Certify maintenance work; load the engine on the test bed and conduct a test run; prepare the engine for despatch and send it to the operator.

The assembled engine is then moved to the test bed. It is not unnatural to find a test bed becoming a bottleneck. Since these are very expensive facilities, an engine shop will rarely own multiple test beds.

The engine is started and run in a simulated environment. All the parameters specified in the EMM are tested again and recorded. The engine is not released until it passes the test.

A 'tested' engine is then prepared for despatch and put in its carriage. The next step is to document the latest status of the engine and despatch it to the operator or owner.

The test bed generates an enormous amount of data, which are mostly relevant to the OEM. However, there are many sets of data points that are captured to verify and declare the warranty and engine's life-related information. The criteria for pass/fail are also sometimes affected by the data generated by the testing process.

3.5 Aircraft components maintenance life cycle

Figure 3.4 shows the life cycle of component maintenance, which is valid for any MRO organisation, in-house or external. This does not include long-term demand and resource planning for the maintenance of components. We assume that at this stage the MRO organisation has

Figure 3.4 Component maintenance life cycle: component overhaul/repair

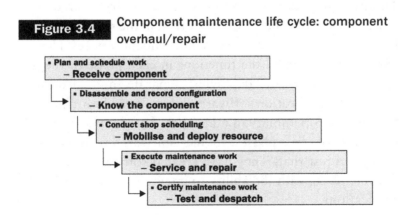

completed the exercise of long-term planning, which we covered in the previous chapter. Moreover, we further assume that the MRO organisation we are referring to is a full service organisation, which may not be the reality in many cases.

In this section we discuss aircraft component maintenance. It is, however, worthwhile reiterating that the maintenance of aircraft components does not fall under the rigorous regime of airworthiness compliance. Almost all the items maintained thus fall under the 'off-wing' category. The cycle and interplay of the functions remain almost same as the aircraft maintenance life cycle, where the component maintenance starts with scheduled or unscheduled tasks and ends with certifying the work.

The functions or phases of the aircraft component maintenance life cycle are:

■ Receive component: plan and schedule work.

■ Know the component: manage aircraft configuration.

■ Mobilise and deploy resources: conduct production planning and control.

■ Servicing and repair: execute maintenance work.

■ Test and despatch: certify maintenance work.

The functions identified as phases, above are discrete and more or less independent of each other, in the sense that these functions can also be defined as independent services. Hence, one can automate all the functions in each phase of the life cycle separately, by deploying separate and dedicated modules of application software. In fact, it can be architected keeping interoperability in mind such that it supports all the functions of the phases. Some software vendors have attempted just that. They have created software to cater to just one phase with the ability to integrate with other MRO applications.

Again, maintaining a component does not require the rigour of aircraft maintenance and does not have to follow the rule of 'if removed – unserviceable'.

However, component maintenance is similar to aircraft maintenance on the following counts:

- The maintenance activity is task oriented and uses AMTOSS Task Cards.

- Major components, like landing gear, require an airworthiness certificate before being despatched after maintenance.

- Component maintenance requires extensive record keeping and preservation of historical information.

- The maintenance is strictly carried out as specified in the OEM manuals (CMM, FIM, etc.).

3.5.1 Receive component

Actions: Plan and schedule work; conduct short-term capacity and resource planning; allocate resources and schedule jobs.

Once the component arrives in the component shop a planning exercise is triggered off and the relevant information pertaining to the life and maintenance history of the component is recorded. If the component is already known the component repair shop will have a record, which is than updated. Otherwise a new set of records is created and subsequently maintained.

Normally a component arrives at the component shop with an unserviceable tag, which has its identification and life information. A defect report may also accompany the component. Based on the information available the component shop creates the work packages consisting of AMTOSS Task Cards.

3.5.2 Know the component

Actions: Disassemble and record configuration; identify record and update component's configuration-related information.

Barring landing gear, the configuration of a component is not necessarily complex. Hence it is rare that a fully fledged configuration management exercise is required for component maintenance.

The record keeping is done for each serialised component and it is ensured that even if the part number changes due to application of a modification, the history is not lost.

3.5.3 Mobilise and deploy resource

Actions: Conduct shop scheduling: plan and manage procurement, inventory and logistics of spare parts and material.

A component repair shop has standard work breakdown structures for the activities that are normally carried out for repairing a component. Based on the work breakdown structures (WBS) and the information available from the Task Cards, the necessary resources are allocated and a requisition for the materials is raised.

3.5.4 Service and repair

Actions: Execute maintenance work; service/clean the components and repair if necessary.

The repair or maintenance work is executed according to the schedule and under the strict guidelines provided in the manuals (CMM, FIM, etc.) from the OEMs.

3.5.5 Test and despatch

Actions: Certify maintenance work; test the components, prepare for despatch and despatch to the operator.

The repaired component is tested as per the CMM. If the component passes the test then a 'serviceable' label is attached and the necessary documents for despatch are prepared.

The component is then despatched to the operator or MRO organisation, whoever is responsible for fitting it to an aircraft or engine.

Normally it is received by the Material Division of the MRO organisation, where it is again inspected.

3.6 Ground support equipment/fleet (GSE/F) maintenance life cycle

Figure 3.5 shows the life cycle of ground support equipment maintenance, which is valid for any MRO organisation, in-house or external.

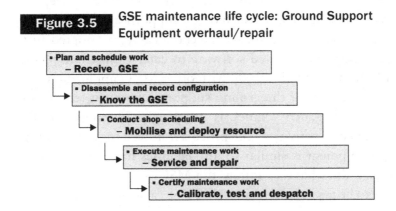

Figure 3.5 GSE maintenance life cycle: Ground Support Equipment overhaul/repair

In this section we discuss ground support equipment/fleet (GSE/F) maintenance. It is, however, worthwhile reiterating that the maintenance of GSE/F does not fall under the rigorous regime of airworthiness compliance. Almost all the items maintained thus fall under the 'off-wing' category. The cycle and interplay of the functions remain almost same as the aircraft maintenance life cycle, where the GSE maintenance starts with scheduled or unscheduled tasks and ends with certifying the work.

The functions or phases of the GSE/F life cycle are:

- Receive GSE/F: plan and schedule work.

- Know the GSE/F: manage GSE/F configuration.

- Mobilise and deploy resources: conduct production planning and control.

- Servicing and repair: execute maintenance work.

- Test and despatch: certify maintenance work.

The functions identified as phases, above, are discrete and more or less independent of each other, in the sense that these functions can also be defined as independent services. Hence, one can automate all the functions in each phase of the life cycle separately, by deploying separate and dedicated modules of application software. In fact, it can be architected keeping interoperability in mind such that it supports all the functions of the phases. Some software vendors have attempted just that. They have created software to cater to just one phase with the ability to integrate with other MRO applications.

The purpose of the ground support equipment (e.g. from ladders to gantry crane) is to provide support to the maintenance activities. The process followed for maintaining this equipment is similar to any facilities maintenance. The process does have to follow the MSG-3 logic but does not have to be task oriented, i.e. AMTOSS cards are not necessary for them. However, in many cases AMTOSS may be applicable. Also, the rigour of maintaining the historical records is not like those for aircraft, engines or components.

Normally the maintenance of GSE/F is outsourced and is not considered to be part of the MRO organisation's core competence. However, in practice many MRO organisations prefer to maintain their GSE/Fs themselves.

3.6.1 Receive GSE/F

Actions: Plan and schedule work; conduct short-term capacity and resource planning; allocate resources and schedule jobs.

A GSE/F may become unfit, either due to breakage or calibration going off track. Once the equipment arrives in the maintenance shop the production planning exercise is triggered off and a job order is created, which contains the activities required to perform maintenance of the GSE/F.

3.6.2 Know the GSE/F

Actions: Disassemble and record configuration; identify record and update GSE/F; configuration-related information.

Since the repair is mainly of 'break–fix' type, it is not necessary to keep strict control of the configuration of the GSE/F. However, it is important to identify and codify the nature of the defect or breakage to correctly perform the repair work.

3.6.3 Mobilise and deploy resource

Actions: Conduct shop scheduling; plan and manage procurement, inventory and logistics of spare parts and material.

A GSE/F repair shop has standard WBS for the activities that are normally carried out for repairing a GSE/F. Based on this WBS and the information available from the OEM manuals, the necessary resources are allocated and requisitions for materiel are raised.

3.6.4 Service and repair

Actions: Execute maintenance work; plan and manage procurement, inventory and logistics of spare parts and material.

The repair work is executed as per the production schedule.

3.6.5 Calibrate, test and despatch

Actions: Certify maintenance work; distribute work, record actual hours and material consumption, support trouble shooting, rescheduling due to non-routine requirements and get the final certification.

Once the GSE/F is repaired, it is calibrated and tested as per the specifications provided by the OEM. If the GSE/F passes the test it is despatched to the MRO organisation, where it is deployed in the facilities or stored in the warehouse.

3.7 Manage materials and logistics

The activity of aircraft maintenance is hugely dependent on the availability of spare parts at the right time and at the right location. The processes used for managing the spares parts such as procurement, receipt/issue, warehousing, inventory control, etc., are similar to any other business.

There is a large body of knowledge available on these topics. However, there are some process areas which are unique to aircraft maintenance and we will discuss them in this section.

3.7.1 Manage demand

In case you happen to be in an aviation MRO organisation's warehouse and overhear conversations between the maintenance engineers and the material managers, you will come to believe that there always is a shortage of parts and you will also feel the panic in the atmosphere. The fact is that there is never a day or an hour when the engineers demand a specific part which is not available in the warehouse!

How does this happen? The root cause of the problem is that there is no method or formula by which an engineer could forecast the failure rate of an item and when and where the failure will happen. There are a lot of statistical algorithms in use but they are all based on historical trend, which is not of much use when a part maybe declared 'unserviceable' even though it has not really failed. Unfortunately the relationship between the component failure rate and removal rate is still not known. It is always assumed to be equal. This in itself makes the forecast for material skewed and a material manager mad.

The demand for spares is triggered by:

- initial provisioning
- scheduled maintenance packages
- unscheduled and ad hoc requisitions
- engineering orders for modifications.

Calculating demand for initial provisioning, scheduled maintenance and engineering orders is straightforward. However, trying to make sense of unscheduled and ad hoc requests is extremely difficult. The only viable strategy is to resort to buffers. But that impacts the bottom line and is also not a panacea.

3.7.2 Strategic sourcing

Strategic sourcing is defined as: 'an institutional procurement process that continuously improves and re-evaluates the purchasing activities of a company. In a production environment, it is often considered one component of supply chain management.'[15]

The scope of strategic sourcing is very limited in the aviation maintenance industry where an aviation MRO organisation has very limited options to choose from.

As discussed elsewhere, the suppliers of aircraft parts are subject to a certification process and they are identified and recommended by the OEMs. The list of suppliers is provided in the IPC (Illustrated Parts Catalogue) and the MRO organisation is restricted to choose from this list. The organisation has the right to change the vendor but again has to stick to the list provided in the IPC.

There are other sources that an aviation MRO organisation taps, namely the second-hand parts market and airlines. However, these sources are not very dependable and may cost more, especially if the seller is an airline.

3.7.3 Manage warehouse

Warehouse Management is defined as: 'Performance of administrative and physical functions associated with storage of goods and materials. These functions include receipt, identification, inspection, verification, putting away, retrieval for issue, etc.'[16]

In order to meet the demand for spares for a moving asset like aircraft, it becomes essential to position them in various different types of locations. Sometimes it is mandated by the regulatory authorities to store the parts in specific locations, which are in the general flight path of an aircraft. There are four types of storage that an aviation MRO organisation normally maintains. These are the base or main warehouse, storage in the hangar bays, storage in the line stations and storage on-board the aircraft. The other types of storage that the organisation manages as well are the IATA Pooled Warehouse and the Bonded Warehouse.

■ *Base or main warehouse:* The main warehouse is the one that keeps most of the inventory of spare parts. It is like any other warehouse with facilities to receive, issue and store parts; which can be bulky, medium or small items.

■ *Storage in hangar bays:* When an aircraft is due for a major check or heavy maintenance, the requirement for spares are called for by the Planning Dept. These identified parts are then issued out against the aircraft and stored near the hangar bay in which the aircraft is due to be serviced. The process is called pre-loading and it is assumed that all the spare parts will be consumed by the check or the maintenance activities. In case some are left over then they are 'returned' to the main warehouse.

■ *Storage in line stations:* These stores support the line maintenance and minor checks. They are generally very small and with limited capacity. Some organisations, when they issue out the parts to the line stations, consider it consumed, even though they might not have been fitted to the aircraft. And some manage them like a mini warehouse, keeping all the transactional records. There is also a possibility that these stores, may become large if some aircraft tend to have a long turnaround time, layover or night halt. The aviation MRO organisation tends to take advantage of the time available and conducts some major defect rectification, which had been deferred, or carries out some checks which are equivalent to A and B Checks.

- *On-board aircraft:* These are also known as 'Fly Away Kits'. Some of the spares on-board are mandated by the regulatory authorities, some are sent with aircraft that are due for a layover and some maintenance work, and some are there just for any contingency. Obviously there is not a log that can be kept on-board the aircraft but some critical as well as generally high consumption parts are also stored there.

- *IATA Pooled Warehouse:* The fact that many airlines use the same airport and similar type of aircraft led to cooperative agreements, which are generally called a 'pooling agreement'. Under these agreements operators pool their spare parts in one specific location and then any of the signatories can use those parts when required. An aviation MRO organisation does not go into such agreements. However, they have access to those parts for an aircraft which belongs to an operator, who has a pooling agreement for the part in the warehouse. The advantage of pooling is that when there is an urgent requirement, the operator does not have to pay a premium price. But the down side is that some participants of the pool would invest but never use those parts. This arrangement works very well where there is very high volume of aircraft traffic.

- *Bonded Warehouse:* 'The building or other secured areas in which dutiable goods may be stored, manipulated, or undergo manufacturing operations without payment of duty.'[17] An aviation MRO organisation maintains and manages one or more bonded warehouse, in order to take advantage of custom duty exemptions and sometimes for some items that the operators might need but the country, where the warehouse is, does not allow them in. These warehouses are mostly located within the main warehouse but with physical segregation and necessary security put in place.

3.7.4 Procurement

Procurement is the acquisition of appropriate goods and/or services at the best possible total cost of ownership to meet the needs of the purchaser in terms of quality and quantity, time, and location. Corporations and public bodies often define processes intended to promote fair and open competition for their business while minimizing exposure to fraud and collusion.[18]

The aircraft maintenance industry was one of the first to adapt the electronic business process for procuring spare parts. The standard for the electronic interchange was developed a few decades ago and was called ATA SPEC 2000. It was designed to cater for TELEX message. Even today the data format in the new standard iSPEC2200, which has absorbed SPEC 2000, is still the same. However, this standard allows an aviation MRO organisation to place orders for spare parts directly and electronically to the vendors and to receive commitments.

In all other aspects the procurement process used by an aviation MRO organisation is similar to any other business.

3.7.5 Manage inventory

Inventory management is primarily about specifying the size and placement of stocked goods. Inventory management is required at different locations within a facility or within multiple locations of a supply network to protect the regular and planned course of production against the random disturbance of running out of materials or goods. The scope of inventory management also concerns the fine lines between replenishment lead time, carrying costs of inventory, asset management, inventory forecasting, inventory valuation, inventory visibility, future inventory price forecasting, physical inventory, available physical space for inventory, quality management, replenishment, returns and defective goods and demand forecasting. Balancing these competing requirements leads to optimal inventory levels, which is an on-going process as the business needs shift and react to the wider environment.[19]

The art and science of inventory control, with its famous reorder points and safety stocks, have been with us since the Industrial Revolution. An aviation MRO organisation also uses the same techniques like any other business, which includes ABC analysis, optimal lot sizing, back order management, replenishment strategies, etc. However, the main difference that comes to mind is the fact that the emphasis on 'safety stock' is much higher in this business than others. Not only does the safety stock need to be managed diligently in the main stocking location, the warehouse, but also how much should be stocked at various line stations? The inventory of spare parts for aircraft maintenance needs to be managed across the network, wherever the organisation performs maintenance activities.

3.7.6 Logistics management

> Business logistics can be defined as 'having the right
> item in the right quantity at the right time at the right
> place for the right price in the right condition to the
> right customer', and is the science of process and
> incorporates all industry sectors.[20]

A commercial aircraft's purpose is to fly from airport to
airport for at least 14 hours a day, and in many of these
airports, called line stations by MRO people, some or other
check, inspection or maintenance can be carried out. As
discussed earlier, each of these line stations also needs tools,
spare parts, consumables and supporting documentation. It
is the logistics arm of the aviation MRO organisation which
ensures that these items are available to the Line Maintenance
engineers and technicians. In addition to that there is a need
to move items from and to the vendors.

Normally the vendors organise the shipment of items
through freight agencies to the warehouse. These movements
are tracked very closely to ensure timely arrival and
availability of the item.

An MRO organisation routinely ships out engines and
components for repairs to third parties and then receives the
repaired items. These movements can become very complex
as they tend to be cross-border and have not only contended
with the complexity of physical movement but also customs
regulations and the large amounts of paperwork related to
these movements.

Another complexity is introduced due to the fact that the
maintenance-related items sometimes require special
packaging and handling. The organisation has to make
sure that the right process and material are provided
for this.

The transportation systems used within airport perimeters are not necessarily general purpose vehicles. The organisation has to ensure that the right kinds of vehicles, as specified by the regulatory authorities, are used for moving the items within an airport.

The other important aspect of logistics management is distribution management. As discussed in this section, the spare parts can belong to one party but be stored in another party's warehouse, which can in turn supply to some other warehouse or location, so the distribution of the same part within the network can become very complex. An aviation MRO organisation uses various channels to achieve efficient distribution of spare parts across its network.

3.8 Manage finance

Managing finance of an MRO organisation is no different from any other organisation. The same principles, policies and procedures apply. Hence, the differences, if any, are more on emphasis rather than different procedure. An MRO organisation also has to follow the company laws and accounting practices of the land they operate from.

In order to keep our focus on the unique aspects of an aviation MRO organisation we will discuss some areas of financial management and accounting, but since there is a large body of knowledge available on this topic, we will not discuss the entire process in this section:

- funds management
- liabilities management
- fixed asset management
- contract valuation and management
- cost accounting.

3.8.1 Funds management

Since the aviation MRO organisations do not own aircrafts, the funding requirement is not as capital intensive as an airline. There are hardly any MRO organisations which are publicly floated. Most of them are either fully owned subsidiaries of the mother airlines or joint ventures.

The aviation MRO organisations, however, require a very high level of cash flow due to pressure of operating expenses, The main are being labour. The labour cost varies period to period and is seasonal, e.g. airlines generally defer their maintenance during summer and winter holidays seasons.

The major factor, which creates unpredictable spikes in the labour cost, is overtime. Since the overtime payments cannot be delayed and could be double or triple the estimated amount, the fund management need to provide for sufficient cushion or access to funds to cater for that.

The other factors that affect the funding of an aviation MRO organisation are customs duty and warranty claims. Since the parts and components can be procured from any exporting country and it is not predictable which one, the amount of funds required for custom duty may fluctuate significantly. Similarly the warranty claims are unpredictable in terms of timing and amount.

Since both factors, overtime and custom duties, are event driven, such as AOG (Aircraft on Ground) or unscheduled maintenance, the funds need to be managed either by providing a cushion or access to funds.

3.8.2 Liability management

An MRO organisation does not own very high value assets like aircraft or engines. However, it stores and consumes high value assets on its premises, which increases its liability and risk. The high value assets that the organisation manages belong to the operators, OEMs and leasing agencies. Some assets are simply stored until used, like pooled items. The loss due to theft, fire and damage during maintenance activities can create significant financial risk to the organisation.

The aviation MRO organisation uses various types of insurance covers to manage the risks and related liabilities.

The other type of liability that an aviation MRO organisation carries is the warranty given to the operators and third parties. The amount of warranties can be very high if looked at cumulatively. And the spread of this liability is also very wide because the warranty covers many aspects of the delivered service.

3.8.3 Fixed asset management

Ideally an aviation MRO organisation should manage only fixed assets, which are categorised, in industry parlance, as plant and buildings. The organisation owns facilities, e.g. hangars, ground support equipment, warehouses, tools and related infrastructure, which are its fixed assets. The total value of these assets is comparatively much lower than an airline. Therefore, for an aviation MRO organisation, fixed asset management is not very complex.

However, since they become custodians of the operators fixed assets, i.e. major components (rotable), they are required to keep track of the value of each of those assets as the value changes based on the type of maintenance done. For instance, if the life of the component is increased or decreased then the application of depreciation will be impacted. Similarly, the scrapping and salvaging of the component affects fixed asset management.

The aviation MRO organisation has to be cognizant of the fact that the industry labels all the rotable and most of the repairable (described elsewhere in this book) as capitalised fixed assets, even though they do not own most of them. It is rare that these items will appear in the ledgers of the organisation, though they will be dealing with significant number of these high value items.

3.8.4 Contract valuation and management

Almost all the revenue generating work that an aviation MRO organisation carries out is based on a contract. The value of the contracts indicates the potential revenue inflow. The organisation keeps track of the contract value but not necessarily quarterly because they are mostly outside the range of share market analysts. However, from a financial performance perspective they evaluate the contracts regularly because the changes to the contracts are frequent, mostly driven by unplanned activities which could be at times more than double the contract value.

3.8.5 Cost accounting

An aviation MRO organisation uses the same cost accounting techniques as any other business. However, the rigour required for collecting the actual spend is very high due to the complexity of the cost drivers of the aircraft maintenance business.

The labour cost fluctuation due to overtime is a major cost driver. This requires a sophisticated cost allocation and collection of actual labour hours spent. The cost accounting complexity is further compounded by the ownership of the parts and components used for maintenance. Some of them are owned and supplied by the operator, some by the leasing companies and some from the IATA pooling.

The other cost drivers such as GSE usage, consumables, tools, etc. are set by the organisation based on the way they are organised. If they own GSEs then they might just allocate an overhead cost to the usage. But if they rent it then the rental will be applied according to the usage.

At the end, the aviation MRO organisation has to come up with a costing model, which is as close to reality as possible.

3.9 Manage human resources

Managing human resources (HR) of an MRO organisation is not much different from any other organisation. The same principles, policies and procedures apply. Hence, the differences, if any, are more on emphasis rather than different procedure. An MRO organisation also has to follow the labour laws and HR practices of the land they operate from. The processes of recruitment, payroll, career management, etc. remain the same as any other business.

In order to keep our focus on the unique aspects of an aviation MRO organisation we will discuss some areas of HR management, since there is a large body of knowledge available on this topic, we will not discuss the entire process of HR management in this section:

- training and learning management
- skills certification management
- time management.

3.9.1 Training and learning management

Every aviation MRO organisation tends to run its own training centre. The reason for this is that the aircraft maintenance skills are not taught in any general education institutions. A few education institutions do teach the theory and practice of equipment maintenance but hardly any of them has invested in specific courses for aircraft maintenance. Then there is a question of certification. The FAA is very clear and concise on the certification process, which prescribe the requisite knowledge and experience but does not specify any methods for achieving them. Hence it is the aviation MRO organisations who become the education faculties and trainers.

The trainees or the apprentices are taken in when they are young, mostly after high school. They are then trained in the maintenance processes and given hands-on training. This leads to getting them a licence. These Licensed Engineers/Technicians are then inducted into the organisation's workforce.

It is interesting to note that the training courses are mostly provided by the OEMs, which are very specific. And these courses lead to what is called Type Certification. In other words, one who undergoes say a specific course on A320 Structures will be certified as an Airframe Technician for A320 aircraft. These type certifications go down to much lower levels and can be as low as component level. The reason for this is that the training is designed more from a safety perspective than from a knowledge perspective. This method also demarcates what a certified or licensed technician can or cannot do.

The aviation MRO organisations also train the fresh graduate engineers but very few of them go for type certification. They are mostly deployed in the planning and engineering side of the business.

Further, this training never stops because the regulatory authorities require re-certification. The organisation thus has to keep track of the validity of every technician's certification and provide them re-training for re-certification. Then, whenever a new aircraft or engine or component is introduced, another spate of training is delivered to ensure that sufficient and certified resources are available for maintenance activities.

3.9.2 Skills certification management

An Aircraft Maintenance Technician refers to an individual who holds a certificate issued by the Federal Aviation Administration; the rules for certification, and for certificate-holders, are detailed in Subpart D of Part 65 of the Federal Aviation Regulations (FARs), which are part of Title 14 of the Code of Federal Regulations. Aircraft Maintenance Technicians (AMTs) inspect and perform or supervise maintenance, preventive maintenance, and alteration of aircraft and aircraft systems.

(AMTs) in Europe must comply with Acceptable Means of Compliance (AMC) Part 66, Certifying Staff, issued by the European Aviation Safety Agency (EASA).

3.9.3 Time management

Since the main cost driver of an aviation MRO organisation is labour, it becomes imperative for the organisation's profitability to collect actual time spent on each maintenance activity. Let us recapitulate from the Aircraft Maintenance Life Cycle that all the aircraft maintenance is task oriented and the AMTOSS Task Card also keep estimates of the time it would take to accomplish the task. The Aircraft

Maintenance Plan is also built based on those, mostly modified, estimates. In line with the same logic, the customer or the operator signs a contract based on those estimates. Therefore, it becomes absolutely essential to ensure that the organisation does not spend more time than estimated at each task level, and if it does it should be able to justify and charge the customer accordingly.

Hence, an aviation MRO organisation not only records the clocking-in and clocking-out of a technician but also the start and end time of each task. The hours spent on each task is signed off by the supervisor and becomes the basis for the cost of labour calculations.

3.10 Manage facilities

The aviation MRO organisation requires facilities to carry out the maintenance work, such as hangars, parking grounds, etc. Specifically from an aircraft maintenance point of view, there are three types of location where the aircraft maintenance work can be carried out:

- in an independent fixed location: hangar;

- at the airport fixed location: airport ramp or apron;

- any parking area within the protected area of an airport.

A 'hangar' by definition is:

> a closed structure to hold aircraft and/or spacecraft in protective storage. Most hangars are built of metal, but other materials such as wood and concrete are also sometimes used. The word 'hangar' comes from a northern French dialect, and means 'cattle pen'.

Hangars protect aircraft from weather and ultraviolet light. Hangars may be used as an enclosed repair shop or, in some cases, an assembly area.[21]

A hangar may have one or more bays or slots. Essentially the objective of facilities management is to provide a working slot to an aircraft for its maintenance as required. The slots are sometime aircraft type dependent. Lately the Airbus A380 forced many aviation MRO organisations to create or modify their facilities to accommodate an aircraft of this size, which is similar in length to a Boeing 747 Jumbo but bigger in wing span and height.

An aviation MRO organisation decides on the number of slots and hangars based on the demand and capacity planning.

A ramp or apron is defined as: 'a part of an airport. It is usually the area where aircraft are parked, unloaded or loaded, refuelled or boarded.'[22]

The choice of location is determined by the type of maintenance to be carried out, i.e. whether it is a daily check, turnaround check, defect rectification or heavy check. The criteria for choosing the hangar is whether it is a major check, heavy maintenance or has large component removals, e.g. engine removal. The ramp and parking lot are opportunistic decisions.

The Maintenance Control Centre (MCC) of the organisation decides and directs the aircraft to the specific facilities.

By definition facility management is: 'an interdisciplinary field primarily devoted to the maintenance and care of commercial or institutional buildings, such as hospitals, hotels, office complexes, arenas, educational or convention centres' [23] . . . and hangars and ramps.

The aviation MRO organisation considers GSE (ground support equipment) as part of it facilities. The overall process of facilities management in an aviation MRO organisation is not much different from any other facilities.

From an aircraft maintenance regulations perspective, the facilities must conform to the standards. Most commercial aircraft hangars throughout the world are designed in accordance with NFPA 409 (NFPA stands for National Fire Protection Association). The other standards that are complied to are for health and safety (OHSA) and environment EN 54 of EPA (Environment Protection Agency).

3.11 Manage continuous improvements

Aviation MRO organisations are guided by the regulatory authorities to continuously improve their maintenance effectiveness. This does not necessarily imply process improvements, but mostly means adherence to recommended processes and compliance to the regulatory guidance. This sometimes does limit an aviation MRO organisation's ability to improve its efficiency. Airworthiness takes precedence.

However, there are two processes which are used to improve the effectiveness of an aviation MRO organisation:

- a reliability program, which is generally based on AC120-17A; and

- CASS (Continuing Analysis and Surveillance System).

3.11.1 Reliability program

The term Reliability-Centred Maintenance (RCM) was first used in public papers authored by Tom Matteson, Stanley Nowlan, Howard Heap, and other senior executives and engineers at United Airlines (UAL) to describe a process used to determine the optimum maintenance requirements for aircraft. ... The first generation of jet aircraft had a crash rate that would be considered highly alarming today, and both the Federal Aviation Administration (FAA) and the airlines' senior management felt strong pressure to improve matters. In the early 1960s, with FAA approval the airlines began to conduct a series of intensive engineering studies on in-service aircraft. The studies proved that the fundamental assumption of design engineers and maintenance planners—that every airplane and every major component in the airplane (such as its engines) had a specific 'lifetime' of reliable service, after which it had to be replaced (or overhauled) in order to prevent failures—was wrong in nearly every specific example in a complex modern jet airliner.[24]

Aviation MRO organisations are encouraged to adapt a Reliability Program. This program is set up by an MPRB (Maintenance Program Review Board) to govern the RCM process. This process is based on recognition of the three principal risks and one auxiliary risk from equipment failures: There are threats to

- safety
- operations
- the maintenance budget
- the environment (auxiliary or special).

RCM offers four principal options among the risk management strategies:

- on-condition maintenance tasks;
- scheduled restoration or discard maintenance tasks;

- failure-finding maintenance tasks; and
- one-time changes to the 'system' (changes to hardware design, to operations, etc.).

A reliability program, based on RCM, requires extensive data collection, analysis and reporting. The detailed specification of RCM can be found in the standard SAE JA1011, Evaluation Criteria for Reliability-Centred Maintenance (RCM) Processes.

3.12 CASS (Continuing Analysis and Surveillance System)

The details of the CASS[24] process are available from the FAA website, therefore we will only discuss the highlights in this section.

The executive summary of the document states:

Since 1964, all air carriers have been required by regulation to conduct continuous evaluations of their maintenance programs. Specifically, Title 14 Code of Federal Regulations (CFR) Parts 121.373 and 135.431 require air carriers to establish a Continuing Analysis and Surveillance System (CASS) to evaluate, analyse, and correct deficiencies in the performance and effectiveness of their inspection and maintenance programs. These regulations do not distinguish between maintenance functions the air carrier accomplishes and those that it contracts out. Nevertheless, the responsibility for CASS remains with the air carrier.

CASS is an air carrier quality assurance system, and must consist of the following functions: surveillance, controls, analysis, corrective action, and follow-up. Together, these functions form a closed-loop system that allows the air carrier to monitor the quality of its maintenance. In a structured and methodical manner, CASS provides air carriers with the necessary information to make decisions and reach their maintenance program objectives. Furthermore, if CASS is used properly, it becomes an inherent part of the air carrier's way of doing business and helps promote a safety culture within the company.[25]

As the name implies, CASS consists of a cyclical process of surveillance, control and analysis. A significant amount of data is collected, analysed and reported from different points consisting of routine, surveillance and various other types of data. The detailed listing of data points is available in the FAA document; however, a sample is shown below.

Routine data:

- aircraft inspections
- scheduled maintenance
- Required Inspection Item (RII) program
- major repairs and alterations, etc.

Surveillance data:

- airworthiness responsibilities
- maintenance manuals
- maintenance organisation and staffing
- maintenance training
- Airworthiness Directive (AD) compliance, etc.

Other data:

- Unscheduled maintenance, including repetitive non-routine maintenance.
- Teardown reports, etc.

The effectiveness of the CASS program is measured against the safety-related metrics. This is done through continuous analysis of the data and audit process.

3.13 Manage environment

An aviation MRO organisation uses a large amount of hazardous and non-hazardous material, which can impact the environment, but in the aviation industry there are strict guidelines on how to manage these materials. The guidelines are prescribed in MSDS (Material Safety Data Sheets) and all the MRO organisations are bound to follow the instructions. This is related to the signage, location, transportation and disposal.

> A material safety data sheet (MSDS) is a form with data regarding the properties of a particular substance. An important component of product stewardship and workplace safety, it is intended to provide workers and emergency personnel with procedures for handling or working with that substance in a safe manner, and includes information such as physical data (melting point, boiling point, flash point, etc.), toxicity, health effects, first aid, reactivity, storage, disposal, protective equipment, and spill-handling procedures. MSDS formats can vary from source to source within a country depending on national requirements.[26]

Currently, the aviation MRO industry has not taken to carbon footprint and trading aspects of managing the environment. However, since the biggest contributor to the environmental issue in an aircraft is the engine, an aviation MRO organisation does make an indirect impact. The better an engine is serviced the less the amount of pollutants emitted by it. Also the cleanliness of an aircraft impacts the fuel consumption. A cleaner wing and fuselage reduces fuel consumption. All these activities are diligently carried out during the maintenance cycles.

3.14 Manage information technology

An aviation MRO organisation consumes and processes a large amount of data. Even though the aircraft maintenance processes were designed to be paper based and manual, with the current volume of data it is simply not feasible to deal with such an amount of complex information on paper-based systems. Therefore, the organisation requires IT management. Traditionally, the MRO IT used to be part of the mother airline's IT organisation. However, even in those times, the corporate IT's constant complaint was that the MRO organisation tended to create its own 'Island of IT'. This was mainly because the need for an information service, technology and its management is very different from an airline.

3.14.1 Information technology management

IT management manages all of the technology resources of the MRO organisation in accord with the company's needs and priorities. Those resources include tangible investments like computer hardware, software, data, networks and data centre facilities, as well as the staff who are hired to maintain them. Managing this responsibility within a company entails many of the basic management functions, such as budgeting, staffing, organising and controlling, plus aspects that are unique to technology, such as change management, software design, network planning, tech support and others.

A typical IT organisation will have project managers, business analysts, architects, systems analysts, specialists, network engineers, hardware operators, developers, programmers, etc. They together manage the Business Applications and Technology departments.

3.14.2 Business applications

An aviation MRO organisation requires a set of applications (software) to enable the business processes as described earlier in this chapter.

The applications that a typical MRO organisation needs are:

- operations management
- content management
- relationship management
- business process management
- asset optimisation
- business intelligence.

Typically, a packaged solution is selected for each application area, which enable the following functions:

1. *Operations management:* This component is used to manage the configuration of all equipment (aircraft/systems/engines/components, etc.). It also provides transactional support for business processes including workflow and management of common operational data. It supports all business functions including engineering, work planning and execution, materials management, HR and finance.

2. *Content management:* Management of maintenance records for legal and contractual compliance, preparation, authoring, assessment, delivery and management of operational and technical publications, manuals, documents and other content including text, graphics, drawings, pictures, videos, etc.

3. *Business process management:* Implementing a systematic approach to monitoring business processes to make an organisation's workflow more effective, more efficient and more capable of adapting to an ever-changing environment. This supports the Continuous Improvement Program.

4. *Asset optimisation:* Planning and modelling for optimising aircraft and complex assembly maintenance scheduling, spare part optimisation and product mix analysis.

5. *Business intelligence:* Reporting and analytical support for technical and financial performance; including condition monitoring and reliability management.

3.14.3 Information technology

Here we elaborate on the technology enablers for the application set that an aviation MRO organisation requires for a platform, which is typically very close to TOGAF III-TRM (Technical Reference Model) (see Figure 3.6).

These are:

- hardware infrastructure
- communications infrastructure
- network services
- operating system services

Figure 3.6 IT architecture

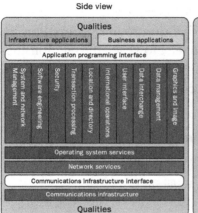
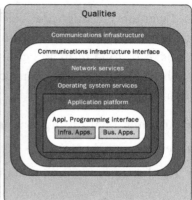

Source: Courtesy of TOGAF.

- application platform
  - graphics and image
  - data management
  - data interchange
  - user interface
  - international operations
  - location and directory
  - transaction processing
  - security
  - software engineering
  - system and network management

- infrastructure applications

- communications infrastructure interface.

A combination of the above provides a platform for enabling the business applications, which in turn enable the business processes.

3.15 Manage external relationships

An MRO organisation needs to interact with its customers and vendors. In their case neither groups are voluminous as in the case of an airline or a retailer. However, the relationship tends to be long term and complex in either case.

There are three sets of processes that an aviation MRO organisation adapts:

■ customer relationship management

■ vendor management

■ regulatory environment management.

3.15.1 Customer relationship management (CRM)

The majority of the customers of an aviation MRO organisation are airlines: major, LCC (low cost carriers), charter, etc. They may also have defence, government agencies, and private owners of aircrafts. However, in most cases the major customer is always the mother airline. Sometimes the ratio of business could be as high as 80:20. Therefore they do not require an extensive customer database or CRM applications. Nevertheless, they need the ability to link up with their customers' systems and also expose their systems to them, especially when they manage their customers' fleets.

3.15.2 Vendor management

There are six types of vendors that an aviation MRO organisation interacts with:

■ OEMs

■ component or parts suppliers

- repair agencies
- IT services
- general services
- commercial items suppliers.

The most critical relationships are with the first three: the OEMs, component and parts suppliers and repair agencies. The MRO organisation's efficiency is directly impacted by the performance of these vendors. The complexity of relationship also increases due to the fact they sometimes store their inventory at the MRO organisation's premises as well as lease them. The interactions are described within the processes in other parts of this book.

3.15.3 Regulatory environment management

An aviation MRO organisation, due to the nature of its activities, operates in a much regulated environment. However, these regulators' main concern is airworthiness. Therefore, the organisation needs to constantly report, give feedback and interact with the regulatory authorities. The local/regional Civil Aviation Authorities are mostly quasi government bodies who oversee and regulate all aspects of civil aviation. The core of the regulation comes from FAA or CAA, but the interaction with local government is similar to any other business organisation.

3.16 Manufacture aircraft parts

Parts Manufacturer Approval (PMA) is a combined design and production approval for modification and replacement parts. It allows a manufacturer to produce and sell these parts for installation on type certificated products. We approve materials, processes, appliances and other parts by other means like a Technical Standards Order (TSO) or in conjunction with a type certificate. Order 8110.42 prescribes the approval procedures for FAA personnel and guides applicants in the approval process.[27]

In earlier times, any parts or components that were intended to be fitted to an aircraft had to be produced by a certified vendor. The certification process is time consuming and expensive. With the improvement in manufacturing technology, FAA decided to relax the certification process and introduced the PMA process. However, it dictated through FAR 21.303(a) that: 'No person may produce a modification or replacement part for sale for installation on a type certificated product unless it is produced pursuant to a Parts Manufacturer Approval (PMA) issued under this subpart.'

An aviation MRO organisation with a component repair shop does have the ability to produce some of the parts themselves, hence there has been significant interest in producing PMA parts by them. They also sometimes supply these manufactured parts to the market. This implies they need to put up manufacturing processes in place, which is different to maintenance processes.

The manufacturing process requires MRP III oriented processes and do not necessarily have to comply with MSG-3 processes. As there is a large body of knowledge on manufacturing processes we will not discuss the topic any further in this section.

3.17 Organisation structure

It is said that an organisation structure follows the strategy of the organisation. Nevertheless, in the world of aviation MRO it is the regulatory body, especially the FAA, which shapes the core of the structure. In Chapter 2 we touched upon the business strategies that an aviation MRO organisation employs. These strategies, formulated by the management, are generally based on the business functions discussed earlier. They are intended to enable the business to run most profitably.

An MRO organisation sets up its overall organisation structure such that it complies with Chapter 4 of AC 120-16D[27] of the FAA and then augments the structure so that it also supports its business strategies. Since the core structure is mandated, it is not surprising to find that the organisation structures of almost all the aviation MRO organisations look very similar. However, the differences, if any, would be in emphasis and designations.

3.17.1 The mandated organisation structure

Since the lives of the general public are involved, an aviation MRO organisation is not free to choose and deploy an organisation structure as it pleases, like other business organisations. However, the way the FAA puts it is that the document AC 120-16D is just advisory. However, non-compliance may result in heavy penalties and censure. Chapter 4 of this document specifically deals with the organisational structure required for maintaining a commercial aircraft.

AC 120-16D Chapter 4 mandates as follows:

You, as an air carrier, are required to have a maintenance organization that is able to perform, supervise, manage, and amend your program, manage and guide your maintenance personnel, and provide the direction necessary to achieve your maintenance program objectives. You are required to include a chart or a description of your maintenance organization in your manual. You can read about all of the maintenance organization requirements in subpart L of part 121, subpart J of part 135, and portions of subpart C of part 119. These organizational regulations apply to your organization as well as any other organization that provides maintenance services for you.[28]

The 'any other organisation' is the aviation MRO organisation, which is the topic of this book.

The FAA provides the specifications of the maintenance organisation and the required management positions. These positions are called Directorate of Maintenance (DOM) and Chief Inspector. In practice though these positions are maintained but are not necessarily called such. The document then elaborates the structure and how it is supposed to deliver the functionality for maintenance. The aviation MRO organisation can name these positions differently and can have different number of positions within the three layer structure that is specified. However, a single person must take the responsibility of DOM, who has the final say in and full accountability for safety and airworthiness of the commercial aircraft that the organisation is carrying out maintenance on.

Since the objective of maintenance is to make and keep the aircraft safe and airworthy, there is a great emphasis on inspections. Therefore, there needs to be a Chief Inspector and an Inspection department. This department must have very clearly defined separation of duties, i.e. it should not be intermingled with the departments which carry out the maintenance programs. Here the strategy to achieve safety is 'Required Inspection Items (RII)'. Hence, the document highlights the significance of the RIIs and how the Inspection department should be organised to manage them.

The FAA has very clearly prescribed the core organisation structure of a commercial aircraft maintenance organisation, which is capable of carrying out the maintenance programs. These maintenance programs are made up of: inspection, overhaul, repair, preservation and replacement of parts.

The advisory document AC 120-16D is very explicit and prescriptive. I would suggest reading through it to appreciate its influence on an aviation MRO organisation's structure. This core structure consists of maintenance operations (heavy and line maintenance), quality assurance, and training departments.

3.17.2 The extended organisation structures

It is interesting to note that Chapter 4 of AC120-16D does prescribe an organisation structure, but this only covers the functionalities of executing the maintenance programs. In other words, it mainly covers the engineering aspect of the MRO organisation. However, an aviation MRO organisation cannot function with just this core organisation structure, as prescribed. It must be noted that this prescribed organisation structure is only applicable to one of the core functional areas, which is aircraft maintenance. The other two core functional areas, the engine and component maintenance areas, are not bound by the FAA regulations.

Within the Aircraft Maintenance functional area, FAA focuses more on the safety related activities, hence it specifies the positions of Chief Inspector and DOM, besides the levels of the organisation. In other words, the regulatory authorities are only interested in knowing who is held responsible, if and when the time comes.

An aviation MRO organisation needs to run a profitable business. Therefore it needs to manage eight supporting functional areas, in addition to the core functional area. These are: Material and Logistics, Finance, Human Resources, Continuous Improvement, Environment, IT and External Relationships. The FAA is silent on these and does not prescribe any organisational structure for them. The logic behind this is that they do not directly affect the airworthiness of a commercial aircraft. However, an aviation MRO organisation has to create an organisational structure to be able to manage these functions, which are absolutely essential for its survival as a business entity. Fortunately, these functions are similar to any other business entities, hence the principle of creating the structure is also similar.

In case the organisation takes up the subsidiary functions GSE and GSF maintenance as well, it extends its organisational structure accordingly. However, these activities are rarely seen to be carried out by an aviation MRO organisation. They are mostly outsourced.

However, recently the trend is to include the PMA parts manufacturing, which requires an addition to the organisation structure.

3.17.3 The evolution of aviation MRO organisation structures

In the initial stages, when there was no concept of an aviation MRO organisation as a separate entity, they were known as M&E (Maintenance and Engineering) Departments or

Divisions of a commercial airline. In those days, and even today with some airlines, the organisation structure consists of Maintenance Operations, Quality Assurance and Training, plus Material Management. All the other functional areas were looked after by the airline corporate office.

Soon the airlines realised that they could do some third-party maintenance. Then they added Sales and Marketing. This introduced the need for keeping track of revenue, cost and profitability. A small finance team was dedicated to the M&E but still reported to the corporate Finance.

Then came the era of independent aviation MRO entities – since they are a business entity of their own, they had to shape the structure such that they could provide the supporting functions as well. That meant that all the eight supporting functions required a department of their own.

The final shape of the organisation structure is now dependent on whether the organisation takes up the subsidiary functions and parts manufacturing as well.

3.18 Summary

The purpose of this chapter was to elaborate on the aircraft maintenance paradigm. In other words the organisation, which has taken up the mantle of maintaining aircraft (especially commercial aircraft), has to follow most of the processes described in this chapter.

We elaborated how the aircraft maintenance life cycle is embedded in the aircraft lifecycle and identified three core functions; two subsidiaries, eight supporting, and one optional function: These are listed below.

The core functions:

- Maintain aircraft
- Maintain engine
- Maintain components.

The subsidiary functions:

- Maintain GSE (Ground Support Equipment)
- Maintain GSF (Ground Support Fleet).

The support functions:

- Manage materials and logistic
- Manage finance
- Manage human resources
- Manage facilities
- Manage continuous improvements
- Manage the environment
- Manage IT
- Manage external relationships.

The optional function:

- Manufacture aircraft parts.

An MRO organisation can have one or all of the three core functions and may also have the subsidiary and optional functions. The optional function is more manufacturing than maintenance. However, whatever the configuration of the organisation, it requires all the support functions to be able to carry out any of the core functions. There are organisation which carry out maintenance of GSE (subsidiary function) or Parts Manufacturing (optional function), independently; but they are not considered aviation MRO organisations on their own.

In the subsequent chapters, first we will discuss how and when the IT industry, especially the software industry, responded to the enablement requirements of an aviation MRO organisation and how the IT landscape evolved. Then we will discuss what are the challenges that the IT industry faces in coming to support the aviation MRO business. And finally we will discuss what the ideal, or the target, is for the IT landscape which an aviation MRO organisation desires.

3.19 Notes

1. Chandogya Upanishad, *Upanisads*, Oxford University Press. Translated from Sanskrit by Patrick Olivelle, 1996
2. *http://www.businessdictionary.com/definition/paradigm.html*
3. APQC – American Productivity and Quality Centre.
4. *http://www.thefreedictionary.com/life+cycle*
5. *http://www.century-of-flight.net/new%20site/frames/myths_frame1.htm*
6. ATA MSG-3; Operator/Manufacturer Scheduled Maintenance Development, Revision 2003.1; Air Transport Association of America, Inc. 1301 Pennsylvania Avenue, NW - Suite 1100 Washington, DC 20004-1707 USA; Copyright © 2003 Air Transport Association of America, Inc.
7. *http://www.businessdictionary.com/definition/certification.html*
8. *http://rgl.faa.gov/Regulatory_and_Guidance_Library/rgMakeModel.nsf/Frameset?OpenPage*
9. *http://www.businessdictionary.com/definition/acceptance.html*
10. *http://en.wikipedia.org/wiki/Technical_documentation*
11. *http://www.businessdictionary.com/definition/phased-introduction.html*
12. *http://www.answers.com/topic/initial-provisioning*
13. *http://www.businessdictionary.com/definition/maintenance.html*
14. *http://www.businessdictionary.com/definition/retirement.html*
15. *http://en.wikipedia.org/wiki/Strategic_sourcing*
16. *http://www.businessdictionary.com/definition/warehousing.html*

17. *http://en.wikipedia.org/wiki/Bonded_warehouse*
18. *http://en.wikipedia.org/wiki/Procurement*
19. *http://en.wikipedia.org/wiki/Inventory*
20. *http://en.wikipedia.org/wiki/Logistics*
21. *http://en.wikipedia.org/wiki/Hangar*
22. *http://en.wikipedia.org/wiki/Airport_ramp*
23. *http://en.wikipedia.org/wiki/Facility_management*
24. *http://en.wikipedia.org/wiki/Reliability_centered_maintenance*
25. Continuing Analysis and Surveillance System (CASS) Description and Models; DOT/FAA/AR-03/70. This report is available at the Federal Aviation Administration William J. Hughes Technical Centre's Full-Text Technical Reports page (actlibrary.tc.faa.gov) in Adobe Acrobat portable document format (PDF).
26. *http://en.wikipedia.org/wiki/Material_safety_data_sheet*
27. *http://www.faa.gov/aircraft/air_cert/design_approvals/pma/*
28. *AC 120-16D Advisory Circular Air Carrier Maintenance Programs, US Dept of Transportation, FAA, Washington DC, USA.*

4

Aviation MRO organisations' challenge to the IT industry

Abstract: This chapter highlights some of the challenges that the unique processes, which are used to maintain a commercial aircraft, pose to the IT industry in general. The chapter provides an overview of challenges faced by the IT industry in providing software and services to an aviation MRO organisation; describes governance issues and maintenance challenges for aircraft; highlights specific uniqueness in the processes; and identifies enablers to meet the challenges posed.

Key words: applications, ERP, systems integration (SI), service oriented architecture, SOA.

Mountains cannot be surmounted except by winding paths.[1]

The summit of airworthiness can be conquered by using various winding paths, which are well marked and enshrined by the regulatory authorities. These routes are well charted in the document AC120-16D by the ATA. These processes have also been discussed in the preceding chapters of this book.

The processes and functions that lead to achieving airworthiness are well prescribed and described. However, in some aspects they tend to be inflexible. Hence, even though the well described processes are music to IT industry ears the inflexibility and prescriptive nature poses severe challenges to them. In this chapter we are going to discuss these challenges posed by the aviation MRO industry.

4.1 Too many aviation MRO standards and lack of them

MSG-3, in principle, is a task oriented maintenance process. This also implies that the aim of these tasks is to ensure that all the Maintenance Significant Items (read components) in an aircraft are serviceable and meet the airworthiness standards. Hence the proper identification of these components or parts is absolutely essential. In order to do so, the ATA specified a standard for aircraft parts, which all the manufacturers are mandated to follow, called ATA100. This has now been merged with other ATA standards and is now part of iSPEC2200. This standard covers the document and electronic data exchange standards as well, which were covered by ATA 2100 and SPEC 2000.

There are three main areas which have been standardised, which cause a lot of grief to a software vendor. These are: Part Identification and Tracking; Document Interchange; and Electronic Data Interchange (EDI).

4.1.1 How do you track and report on the movements of a chameleon?

The part numbering standard is quite rigid in its specification. However, it allows each vendor to use their own part number for exactly the same part hence the same part may be identified, and legally so, by different manufacturers by a different part number.

It works like this. Boeing will design and specify a part and assign a part number to it. This part may or may not be produced by Boeing itself and may be produced by more than one vendor. Now, each vendor is obliged to follow the ATA standards but not obliged to use the same part number that either Boeing or any other vendor has specified. So the same part can be identified by more than one part number.

These part numbers are called manufacturer's part number for the reason described above. Therefore the only way a part can be identified uniquely is by adding the vendor identification number to it. But the good thing is that all this information appears in the Illustrated Parts Catalog (IPC) at the same place.

This situation requires every instance of an aviation MRO application to create and maintain its own part numbering scheme. This solution does work for and within the company but this identifier becomes useless for any communication with the outside world: other airlines, MRO organisations, OEMs, etc. The external world only understands the manufacturing part number.

To add to this, the part number can also change if there is a modification carried out. And, they can be configured differently so they become interchangeable. So a part can be an alternate to another part or interchangeable with another part; can be superseded or made obsolete. In a nutshell, an aircraft part number is like a chameleon but the law requires that every movement of this chameleon is tracked and reported meticulously, consistently, and regularly; from its birth to death.

Further, new technology is acerbating the situation. The new aircraft nowadays carry a lot of software, which are treated as maintenance significant items because they affect the airworthiness of the aircraft. Hence, they need to be numbered and identified according to the ATA standards and tracked like any other item. Since these parts do not have any physical presence, they pose a unique problem when it comes to their removal and installation.

In the IT industry, no generic Enterprise Resource Planning (ERP) software is geared for tracking this chameleon. However, all the specialised and bespoke applications track and report very competently.

4.1.2 The aviation industry did not like HTML

The aviation MRO industry requires tonnes of documentation, if we ever try to print and measure them. Goldfarb came up with the idea of SGML, which the DoD adapted, to solve the problem of how to deliver the loads of documentation required to maintain an aircraft. The airline industry came up with its own standard ATA 2100, while W3C was devising HTML. However, HTML did not fully meet aviation MROs' requirements so the ATA 2100 won; and all the documentation generated by the OEMs were created using this standard for the purpose of document exchange.

This implied that anyone who wanted to use the electronic documentation created by the OEMs needed to write their own programs or buy a software package, which could access and manage this information. In affect, the aviation MRO industry had thrown open a new standard for document management, which was not used by any other industry. And these documents are not simple, they include all types of drawings, including wiring diagrams.

4.1.3 An early adapter of B2B

The aviation industry went into business-to-business (B2B) electronic transaction, much before it came into vogue. They were, as far as I know, the earliest adapter to this concept, and developed a standard to accomplish this activity. It was SPEC 2000. This standard allowed an airline or an MRO organisation to communicate with the vendors electronically for the procurement of spare parts. An MRO organisation can directly place an order on a vendor's or OEM's computer system. In fact the entire order processing cycle can be accomplished electronically.

However, in order to keep the technology available then (there was no internet when this standard was developed), the facility was provided that it could be operated using TELEX machines. So even today the message format used is like a TELEX message and one has to be registered for SPEC 2000 and obtain a special address to be able to conduct B2B transactions.

4.1.4 The supply chain is tight and incestuous

The fear of dropping out of the sky and obliterating innocent lives has forced the industry to limit the number of vendors and suppliers who can produce and supply any part which are meant to be fitted to an aircraft. This makes the supply chain very tight and inflexible. The suppliers are fixed and they operate out of fixed locations. Similarly, the OEMs are also very limited. In fact they can be counted on fingers: Boeing, Airbus, Rolls Royce, General Electric, Pratt & Whitney, etc. and some more. These suppliers are specified in the IPC for each part.

However, the regulations stop there. Beyond that the industry is free to manage its supply chain. Moreover, the regulators have loosened their grip further by introducing PMA certification, under which an MRO organisation can also produce some select parts.

The supply chain management within the aviation MRO industry is in fact left to operate on its own and very little standardisation is seen across the industry.

Any suppliers who can produce and supply these parts have to be certified. And they have to keep their certification current and valid. This exercise is not cheap and adds to the price of the part. The other aspect is their production capacity and distribution capacity, which cannot be limitless. Hence the scope for optimising the supply chain by applying techniques like just-in-time becomes limited. In other words, general supply chain concepts either become an overkill or ineffective. The effectiveness of the supply chain for an aviation MRO organisation is extremely dependent on its geographic location, its relationships with the suppliers and the skills of the people managing the supply chain.

An aircraft is built of parts mainly supplied by the following types of certified suppliers:

- aircraft manufacturer
- aircraft engine manufacturer
- aircraft component manufacture.

Essentially they are the OEMs. They provide the parts for the new aircraft and continue to supply spare parts for its maintenance as well. When a new aircraft is delivered to an operator, it comprises of parts procured by the aircraft manufacturer and also parts procured by the operator. This type of part is known as BFE (Buyer Furnished Equipment). Also some of the parts are provided by the suppliers who are not necessarily the original aircraft manufacturer, which are called SFE (Supplier Furnished Equipment).

Once the aircraft is in service, in addition to the types mentioned previously, there are some different ways of procuring the spare parts, which further complicate the supply chain processes. These are:

- pooled items
- borrowed items
- robbed items
- from service agencies
- other airlines and MRO organisations.

The primary market, one can say, is where the parts are procured directly from the suppliers who are also manufacturers. However, there is a secondary market where the airlines and aviation MRO organisations buy and sell parts to each other. This is done directly or through an intermediary like ILC. The secondary market is very unreliable and volatile. The pricing, available quantities and shipping methods and processes do not lend themselves to optimisation. The transactions are reactive, instead of proactive. In this market, generally, the supplying party tries to dump its obsolete and overstocked parts at best possible price. And the buying party goes looking for parts in this market when the primary market is unable to satisfy its demand. This market is analogous to a garage sale.

The secondary market has arisen mainly due to the inherent inefficiencies in the supply chain of this industry.

In addition to these primary and secondary markets, the airlines and aviation MRO organisations indulge in the practice of loan and borrow. In this scenario the buying party, having failed in procuring the part from either the primary or secondary market, tries to beg or borrow from another airline or MRO organisation. The pricing again in these transactions are arbitrary and the supply depends on the goodwill and relationships between the involved parties. There is another issue which often crops up: tracking the life of the part. Unlike the obligations of the sellers in the primary and secondary markets to provide authenticated and certified life data of the part, the borrowing party does not have such strict contractual obligations and may not have any mature process to provide the life data accurately. In addition to that this type of transaction creates problems for financial accounting.

It is interesting to note that the engine manufacturers have created a big business around this concept. They position serviceable engines at various locations and loan them to needy airlines at a price, which is never cheap.

However, this is not the end of peculiarities of the supply chain in the aviation MRO industry. There are two more parts procurement methods, which add to the complexities. One is that there are new agencies/brokers, who are offering the part procurement services. This practice is still in its infancy. The agencies offer to hold an inventory of parts for the MRO organisations. However, they are not necessarily distributors of parts in the real sense. They are not even well defined channels, as understood from the supply chain perspective. These agencies are generally service providers or extensions of OEMs, who contract with the aviation MRO organisations to provide the additional service of parts procurement. Unfortunately there are no standard procedures for these activities. Hence, each and every contract is tailored specifically for every instance.

Last but not the least is what is affectionately called 'robbery' in the industry. This happens when an aircraft, which is on the ramp and in operation, needs a part and there is an aircraft under maintenance in the hangar, which has the part fitted and is serviceable. The responsible engineer then has no compunction in 'robbing' the part from the aircraft being maintained and fitting it to the needy aircraft. As the name implies, it is illegal because any part which is removed from an aircraft, under whatever circumstances, is rendered unserviceable. Hence, before being fitted to another or same aircraft, the part needs to go through the process of making it serviceable. Obviously, this activity of 'robbery' does not respect the 'law' and normally happens at night, justifying the name. The implications of the activity are obviously loss of tracking data. An aviation MRO organisation ends up doing a lot of reconciliation of life data, the most important measure of airworthiness.

And then there is life after death. There are many mothballed aircraft, which are parked in close proximity to an aviation MRO organisation. In this case 'cannibalisation' takes place. However, in this case the cannibalised part goes through a rigorous inspection process before being fitted to an active aircraft.

4.2 The conundrum of ownership

Let us now pop the question: Who owns the parts fitted to an aircraft: the airline, MRO organisation, OEM, service provider or a leasing agency?

The answer is never absolutely clear. This situation is muddled further with the fact that an aircraft cannot only be owned or leased, but can also contain parts procured and supplied by the several MRO organisations, who maintain the aircraft.

A flying commercial aircraft is hardly ever fully owned by a single entity from its components perspective. However, from the aircraft's maintenance perspective it becomes more complex because these ownerships also impose some constraints, which interfere in the fitment and removal of parts.

The biggest issue raised by varied ownerships of the parts fitted to an aircraft is in the area of finance, which includes valuation of a component and calculation of maintenance cost. The problem is further augmented by the requirement sometimes imposed by the owners to manage the inventory of these parts separately. Therefore, the same item may end up getting different treatment in terms of procurement, storage, usage and scrapping.

4.3 Forecasting: does the industry have a crystal ball?

We have discussed the forecasting requirements in the earlier chapters. However, it is worth elaborating them further to enunciate the issues related to forecasting in the aviation MRO industry.

An aviation MRO organisation needs to be able to forecast the following:

- How many aircraft is it going to maintain or service over a long-term planning horizon? This will help in planning the resources required to deliver the services.

- Which components, especially the airworthiness related components, are likely to fail and when? This will help in planning the procurement of parts and maintenance activities required in the short term hangar and slot planning.

- Which spare parts and how many of them should be stocked at which location by the aviation MRO organisation?

The crux of the maintenance forecasting is based on the propensity for failure of a critical component, which could render the aircraft unserviceable. As we know, all components have a life span and they loose their life as they face the environment and regular impacts, accidental or otherwise, from various sources. This decrement of serviceability, which happens as the component slides away from its designed purpose, actually takes place very slowly. In other words it creeps up, sometimes seemingly randomly and sometimes predictably. The main reasons for the gradual reduction of serviceability are: creep, fatigue, and wear and tear.

The R&D world and the academia is buzzing with efforts to model these phenomenon but they have not yet percolated to the practical levels to be of real use to an aviation MRO organisation. With this as the context, we will now discus the various requirements or conundrums of forecasting that aviation MROs have been trying to deal with.

4.3.1 Long-term forecasting

An aviation MRO organisation invests in large and capital intensive facilities and equipment such as hangars and GSE. The hangars and the bays are designed to service various types of aircraft, however, they have limitations too. Some of them can only handle specific types and size of aircraft, hence the organisation has to make well informed decisions before investing in the facilities of certain shape and size.

In order to accurately forecast what facility will be required in the longer term, a significant amount of data and complex algorithms are required. It was easier when the requirements were tied to one mother airline, whose strategic plan for fleet expansion was known. This at least allowed the MRO organisation to aim for the number of bays and hangars based on the types and number of aircraft that the airline was planning to operate over a 10 to 15 year period. However, the frequency of maintenance, though exquisitely planned and declared in the Maintenance Program, required a planning exercise based on complex forecasting models. Having said that, it must be noted that an independent aviation MRO organisation does not have the luxury of such planning data because they cater for more than one airline and the airline's growth plan may not be aligned to the MRO organisation's growth plan or facilities requirements.

When an MRO organisation was (and in some cases is) servicing only the mother airline then the facility planning was reactive, in the sense that the facilities were planned for based on the mother airline's fleet plans. However, once an MRO organisation becomes independent, it has to pro-actively plan for what types of service offerings for which types of aircraft it will go for. It has to strategise and plan for servicing the market rather than a specific airline. This planning requires market research, analysis and long-term demand forecasting.

4.3.2 Forecasting component failures

A component inevitably fails under the stress and strain of mechanical wear and tear, notwithstanding the brunt of variegating and treacherous environmental elements: wind, storm, hail, extreme temperatures, thunder, lightning and so forth. This is predictable. However, what is in question is the unpredictable. There are various mathematical models which have been developed and in some instances they are pretty accurate in their prediction. This works very well for the non-MSI (Maintenance Significant Items). However, for the MSI, it is important that they are serviced or inspected before they fail. Since the airworthiness of an aircraft is dependent on the MSI, it is important to predict when they should be inspected or serviced.

In spite of the fact that the entire MRO operation is dependent on one extremely important parameter, the rate of failure, there is not enough support from the IT industry in this respect.

Let us look at the complexity of this issue. A component or a part is designed and manufactured with a clear intention of not failing for a specific number of hours, cycles, flights, landings, etc. Based on the knowledge of the material and the dynamics of its stress and strain during the aircraft operation, the life of the component is predicted. The environmental factors are also taken into account in predicting or forecasting the life of the component. However, once the component goes into operation the predictability reduces considerably. This happens because measurement of the effect of each flight on the component is almost impossible.

It is a regulatory requirement to track the failure and failure rates of the components. Each actual failure is considered an incident and the goal of an MRO organisation is always to minimise or to banish all together these incidents. In pursuit of this goal, the components are either inspected or serviced well ahead of their points of failure. This introduces a bias.

4.3.3 Forecasting for material . . . how much and where to store them?

At the time of induction of an aircraft, an airline goes through the process of Initial Provisioning – described in the earlier chapters. The OEMs provide a list of spares with recommended quantities and the vendors who could supply these. The recommended quantities are not necessarily of but mostly skewed towards, the OEMs' view of the spares requirement. An airline always goes through the exercise to ensure that they stock an optimum quantity of the spares and also take advantage of the commonality and rationalisation of the spares across the fleet. The OEM's recommendation is generally based on each aircraft that is delivered, with a notional guidance for rationalisation as the fleet size grows.

Added to this is the fact that the regulatory authorities also impose some constraints from the safety perspective. Moreover, some items have to be stored in certain locations based on the flight path and routes of the aircraft. These factors may contribute towards higher inventory levels, especially once the aircraft goes into service and regular maintenance is conducted.

The aviation MRO organisations continuously try to optimise the inventory, but currently it is mostly based on experience and guesswork.

4.4 On-wing vs off-wing: the life value cascades

It may sound obvious, but the key difference between regular plant or machinery maintenance and aircraft maintenance is the fact that the execution and recording of the maintenance activities are different when the maintenance is done on-wing or when it is done off-wing.

This subject has been discussed earlier as well. However, in order to emphasise the affect of these two kinds of maintenance activities on the IT solutions, we shall discuss these again.

The terms 'on-wing' and 'off-wing' do not necessarily imply that the maintenance activities are done on or off wing. What is implied is that once the component is taken off-wing the meter recording its life stops ticking. In other words, an engine or a component when removed from an aircraft must record the life, in terms of flying hours, landings and cycles of the aircraft. This information needs to cascade down to all the maintenance and inspection significant assemblies and sub-assemblies of the engine and the component. It also needs to record which aircraft the component has been taken from.

However, once the engine or the component is off-wing then any assemblies, sub-assemblies and parts removed from them do not require any life data to be additionally recorded. Therefore, it is the treatment of the life data that differentiates the on-wing and off-wing maintenance activities.

Nonetheless, the on-wing maintenance activities are rigorously regulated as they directly affect the airworthiness and safety of the aircraft, hence it is imperative that the life data is diligently recorded and the related history of the maintenance kept until the aircraft goes out of service.

The life of a component or a part is calculated based on the time it has been on-wing of any aircraft. The moment the item is fitted to an aircraft its meter or clock starts ticking and continues until it is removed from the aircraft. This life is to be recorded if the item is maintenance or inspection significant, i.e. it affects the airworthiness of an aircraft. The standard method is to record this time at the removal and installation of the component to an aircraft. These two events define the start and end of each cycle of time recording. In other words, the life of an item is the cumulative time spent between each installation and removal from an aircraft. This means that each event of removal and installation needs to be accurately recorded.

So far so good, however, as long as the item is an aggregate or main unit, i.e. it is not a sub-assembly, the recording of the time or any other life-related parameters is relatively simple. But when the item is a sub-assembly then the recording could become tricky because the levels of sub assembly structure could be as many as seven or even more. Each sub-assembly, that can be removed from or installed in its parent assembly, can have its own unique and distinct life, which needs to be recorded and its history maintained for regulatory purposes.

What this means for an IT application is that it must be able to cascade the life values to the lowest level of an assembly hierarchy, if the item fitted at that level is in any way defined as being related to airworthiness of an aircraft. So, in simple words, when the component or the aggregate is removed from an aircraft, the time it spent on the aircraft is cascaded on to every sub-assembly of the component, with the constraint in mind that they are recorded accurately and quickly.

This life value of an item can be TSN (time since new), TSO (time since overhauled), landings, cycles, etc. The whole idea is that life of an item must be recorded as it works with aircraft from its inception to its demise. This life value is the key to its serviceability or airworthiness.

4.5 MRP-3 vs MSG-3: the ERP paradigm does not work

In this book we have been referring to MSG-3 for quite a while. The reason for this is that the maintenance processes defined for an aircraft are based on the concepts defined in MSG-3. These processes are then more or less institutionalised by the FAA and related regulatory authorities, hence the processes in themselves are akin to statutes. They cannot be violated. One can tinker with them for improving efficiency but cannot avoid following most of them. Thus the aircraft maintenance processes are prescriptive.

Similarly, when MRP-3 was defined for manufacturing industry, the idea was that all the manufacturing could be done efficiently using these processes, which were called best practices. These best practices, however, are not prescriptive, in the sense that if a manufacturing organisation does not follow some or all of the best practices, it is not subject to legal or regulatory actions. Whereas an aviation MRO organisation, if it does not follow the prescribed processes or practices, is deemed to be in violation of law and can incur severe punitive damages including loss of licence.

This implies that an aviation MRO organisation is not allowed to change its maintenance processes, especially from an engineering point of view, whether it is supposedly efficient or not.

It so happened that MRP-3 was extensively used by the IT industry, especially SAP, to develop application, which used the best practices for manufacturing. These applications essentially require that the organisations who implement these ERP systems must change their current processes to suit the best practices embedded in the application.

Unfortunately, the aircraft maintenance processes do not fully comply with the ERP best practices. When an ERP system is implemented by an aviation MRO, especially in the engineering area, the application is forced to undergo significant changes. But when implemented by an organisation, it is bound by the regulations imposed for airworthiness requirements.

The inflexibility of an aviation MRO organisation becomes a singular impediment in implementing an ERP application. However, it is worth noting that since the regulatory authorities are almost silent in the area of material management an application based on ERP concepts is much easier to implement with all its best practices. In the areas of Material Management and Warehousing an aviation MRO organisation can easily change its processes to suit the best practices of an ERP application.

4.6 Every maintenance activity has to have a Task Card

When MSG-3 was published, the major change, which was also a paradigm shift, was that it recommended task-oriented maintenance, instead of process oriented maintenance. This effectively changed the way the maintenance planning was done under MSG-2. It made the lettered checks obsolete and introduced phased maintenance based on specific dates and a set of tasks. In theory, the required maintenance work is dependent on the number of task, that need be to performed on a specific date or duration.

This is an elegant concept, unique to aircraft maintenance.

Aircraft maintenance consists of maintenance procedures, maintenance events, item discrepancy and fault isolation. These are documented using the concept of ATA classifications, which are standardised. We will discuss this further.

4.6.1 It is a Task Card and NOT a work order or job order

In the world of maintenance, the most important piece of document is a work order or a job order. This document comprises of all the specific information required to carry out the maintenance work. However, in the world of aircraft maintenance, this is not true. This is because non-aviation maintenance is process oriented, whereas aviation maintenance is task oriented.

Aviation maintenance uses the standard AMTOSS (Aircraft Maintenance and Task Oriented Support System), as well as JEMTOSS (for Jet Engines) and MTOSS. These are specific task numbers, intelligent numbers that apply to each Task Card, defining them uniquely. Therefore, a work package in aviation maintenance is a set of Task Cards with specific numbers, not just line activities of a job order or work order.

Currently, the Maintenance Procedures Working Group of MSG is working on doing away with AMTOSS/JEMTOSS/ MTOSS numbering scheme and replacing them with a different structure. The manufacturers were running out of numbers for the AMTOSS/JEMTOSS/MTOSS code, because it was intelligent. MPWG therefore decided to identify tasks and subtasks by a unique non-intelligent number. The elements of the AMTOSS code are still there and can be retrieved but they are no longer used as the identifier. However, the maintenance process will remain task oriented and will not conform to the work order or job order oriented processes.

It must be noted that each task is created by the OEMs and aviation MRO organisations use them as is, along with some of their own specific Task Cards.

4.7 Many systems to integrate

The aviation MRO operations are complex and require an extensive amount of data. In order to support the functionality, it has been noted that a single application is not enough. In some cases I have seen more than 300 small to large applications being used. A lot of these applications are due to piecemeal developments and lack of an all-encompassing software in the market. However, there is a significant amount of data that is required to support the MRO organisation, which need to be drawn from applications, which are out of the scope of MRO functionality, e.g. Finance, HR, Training, etc.

If we assume that there is core MRO application, which covers all the required MRO functionality then the information required by this core application will be from internal support applications: HR, Finance, Training, Procurement, etc; and from external systems belonging to: OEMs, vendors, customers, regulatory authorities, Customs, ACARS, etc.

A representative integration requirement is shown in Figure 4.1 This diagram is a simplified conceptual depiction of integration between key internal applications, a typical requirement for system integration. Though simplistic, the applications pose considerable technical challenges for actual implementation.

There is another view of the integration requirements, which is shown from a functional perspective in Figure 4.2.

The external systems include the applications on-board: the Health Monitoring applications for both aircraft and engine; Electronic Flight bags, etc. They need to be integrated with the core MRO application. The data which flows in from a modern aircraft are:

■ electronic log books: pilot, cabin, defect and technical;

■ faults and conditions.

The other types of external systems are: documentation systems from OEMs, which provide online access to documentation; the vendors' systems for B2B operations; customers' maintenance planning applications; miscellaneous web-based applications, which provide various types of services; etc. The types of interactions and data are:

■ Documentation: technical, flight, weather, etc.

■ SPEC 2000: B2B part procurement.

■ Customer interactions: work packages, progress reporting, etc.

■ Regulatory: mandatory reporting, service bulletins, airworthiness directives, etc.

Overall, a wide set of applications must be integrated in such a way that by interoperating amongst themselves they

Figure 4.1 MRO system integration

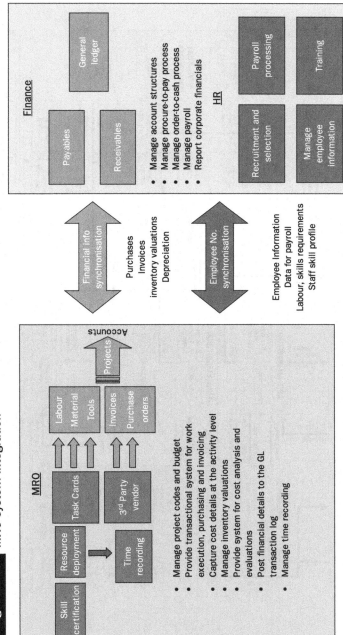

Figure 4.2 MRO functional integration

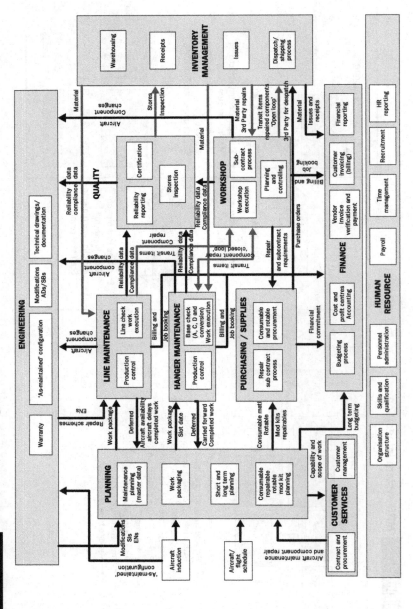

provide all the information and services required by an aviation MRO organisation.

4.8 Summary

In this chapter we discussed the unique challenges that aviation MRO functions and processes pose for the IT industry. We did not go into deep dive, which is beyond the scope of this book. The intent was to tease out the uniqueness and esoteric nature of aviation MROs and not go into defining the detailed requirements for MRO applications. In doing so we covered the following six specific areas, which are not exhaustive:

1. aviation MRO standards
2. ownership
3. forecasting
4. on-wing v. off-wing
5. task-oriented maintenance
6. system integration.

The idea of pointing out these specific areas is to highlight the pain points felt by most of aviation MRO organisations, where they are looking for some support and innovative products from the IT industry.

In the next chapter we discuss what the IT industry has done in the general area of software and application provisioning for aviation MROs.

4.9 Note

1. Johann Wolfgang von Goethe.

5

The IT industry responds

Abstract: This chapter highlights how the IT industry responded to the needs of the aviation MRO. It takes the reader on a short journey through time when the IT industry matured from being just middleware provider to application provider. The chapter describes the enablers and cutting edge technologies that were deployed to support the aircraft maintenance processes and subsequent business. It explains how and why the IT industry initially ignored this domain and then came to partake, but is still trying to catch up.

Key words: Mainframe, MAXI-MERLIN, SCEPTRE, SABRE, client/server.

If in this world a man comes to know it, to him belongs the real. If in this world a man does not know it, great is his destruction.[1]

It might have been a great coincidence, or the mighty convergence of technologies, when the aviation MRO and IT industries arrived on the scene almost at the same time. MSG-1 followed the cutting edge technology of Boeing 747, which opened the skies to the world; and IT arrived riding the wave of IBM's System 360, a consolidation of computing power open for any industry to use as a business enabler.

The seeds of both industries were sown in the 1970s and germinated and proliferated at a breakneck speed. The airline industry took to IT like a fish to water. The famous SABRE online system was developed at American Airlines with the help of IBM. Similarly a number of solid MRO (they were called M&E systems then) were introduced very successfully. They were all bespoke and developed by the airlines themselves. However, when the independent IT software companies arrived on the world stage, they did not think that the aviation MRO was worth investing in. They ultimately came back and did, but it took them about twenty years to do so. There was a period when the IT industry did not even consider aviation MRO, because in their perception it was not a viable market. We will discuss this further in this chapter.

5.1 The aviation MRO business and information technology

In MRP (Manufacturing Resource Planning), the concepts and the processes did not take into account that computers would end up being its best enabler. Similarly, when the standard processes for aircraft maintenance were developed the team involved did not think of IT. This was the 1970s and the computers of that era were in their infancy. They had very recently moved from football field sized research labs to corporate buildings. System 360 from IBM was leading the assault in the business world. However, when the business processes for aircraft maintenance were ready to be implemented using paper-based systems first, it became obvious that a computer would be a better enabler for these prescriptive processes.

Then American Airlines, along with IBM, launched itself into the brave new world of mainframe computing by creating the SABRE online reservation system. The airlines were in the regulatory environment, which could also be viewed as a monopolistic or cartel type environment. The cost was of no concern – to some extent. A large airline could spend on large innovative IT without much concern for cost.

In the 1970s an MRO organisation, as an independent entity, did not exist. It was always a M & E division of the airline. The M&E division would be a cost or burden centre. However, it wielded immense power to influence the airline management, to the extent that they could get funding for anything they wanted. They had the secret weapon: AOG (Aircraft on Ground). The longer an aircraft stays on the ground the greater the loss for the airline. Obviously, anything that could reduce the on-ground time, the airlines would invest in. Also the airline did not mind because the industry was not yet hit by de-regulation. There was enough financial cushion to be able to indulge in risky investments, whether IT or something else.

That beckoned the era of the bespoke systems. There was the wonderful System 370, evolved out of 360, and a powerful database system IMS with, programming languages COBOL and PL1. It was time for great M&E applications.

5.2 The era of bespoke systems

For the sake of consistency, we will continue to refer to M&E as aviation MRO. As discussed earlier, the aviation MRO business generates a relatively low volume of transactions but these transactions are significantly complex. This means that each transaction that is carried out requires a considerable amount of complex processing. However, the best part is that the requirements, especially for the engineering aspect of the aviation MRO, were standardised and explicit. Therefore the performance, i.e. ability to run complex transactions, was the main driver for an application that could enable aviation MRO processes.

Many in-house bespoke applications were developed during that decade. Almost all the major airlines, American Airlines, British Airways, USAir, Alitalia, etc., developed their own MRO applications. However, only a few are noteworthy. Then there were three waves of application development between the mid-1970s and the early 1990s. These waves were triggered by the technologies such as database and programming languages being developed. We will discuss these waves shortly.

Just to recap on the basics of IT, an application that is capable of enabling a business processes requires, at the minimum, a set of tools: a programming language, an operating system, a database, a communications system, a network and computers. And they should be able to meet the performance needs. In 1970s the necessary IT tools arrived. They were, COBOL and PL1 for programming language; MVS and DOS/VSE for operating systems; IMS DB for database; IMS DC and CICS for communications systems; SNA for networks; the famous green screen 3270 terminals or workstations; and System 370 for computing.

Before we jump into the cycle of application development, let us review the building blocks which were used to develop the aviation MRO applications.

5.2.1 Programming language

Programming language is a 'coded language used by programmers to write instructions that a computer can understand to do what the programmer (or the computer user) wants'.

Before the advent of business oriented programming languages such as COBOL and PL1, writing a solution for an aviation MRO was almost impossible.

COBOL (Common Business Oriented Language), developed in 1959, is one of the earliest high-level programming languages. It was guided and endorsed by the Conference on Data Systems Languages (CODASYL) and is said to have been developed within a six month period, and yet is still in use over 50 years later.

PL/1 was developed by IBM in the mid-1960s, and was originally named NPL (New Programming Language). Later, the name was changed to PL/1. Because the language was developed at IBM's Hursley Laboratories in the United Kingdom, the name was later changed from NPL to PL/1 to avoid confusion with the National Physical Laboratory in England. PL/1, also known as THE programming language, was intended as an all-purpose language to combine the scientific abilities of FORTRAN with the business capabilities of COBOL, plus additional facilities for systems programming.

Until the time this new language was developed, all previous languages had focused on one particular area of application, such as science, artificial intelligence, or business. PL/1 was not designed to be used in the same way. It was the first large scale attempt to design a language that could be used in a variety of application areas.

Both COBOL and PL/1 suited the development of aviation MRO applications. They provided the platform for coding complex business functions in a reasonable time frame. Some airlines chose COBOL and some PL/1, but either way they came up with very robust and function-rich applications.

143

5.2.2 *Operating system*

The operating system is the software that controls the allocation and usage of hardware resources such as memory, CPU time, disk space, and input and output devices. This is the heart of a computer system. The operating system makes the hardware usable, the hardware provides 'raw computing power'. The operating system makes the computing power conveniently available to users, by managing the hardware carefully to achieve good performance.

In 1967 IBM released OS/360, which could run on System/360. It was so named because it was aimed at a full circle of customers, from business to science – customers who did a lot of mathematical calculations as well as those who did simpler arithmetic on large sets of data. OS/360 was developed at a cost of approximately half a billion dollars, deploying more than 1000 people including programmers, technical writers, analysts, secretaries and assistants – and all together some 5000 staff-years went into design, construction and documentation. OS/360 was perhaps the biggest and the most complex programs that have ever been attempted. However, it facilitated timesharing and multi-programming in a big way.

In 1970 a steady workhorse of an operating system that could meet demands for high and reliable performance was developed by IBM, using OS/360 as a stepping stone. It was called MVS (Multiple Virtual Storage), named after the technique it uses to manage memory. It lets any user work with huge amounts of memory at once, making MVS ideal for batch processing.

It is evident that the development of an aviation MRO application needed an operating system like MVS.

5.2.3 Database and Database Management System

A database is a collection of information and a means to manipulate data in a useful way, which must provide proper storage for large amounts of data, easy and fast access and facilitate the processing of data. The Database Management System (DBMS) is a set of software that is used to define, store, manipulate and control the data in a database.

In 1960, two main DBMS were developed: network DB (CODASYL) and hierarchical (IMS) DB. There was a third DBMS, developed in Germany, called ADABAS (Adaptable Database System) in the same period which also played an important role in the development of aviation MRO applications.

American Rockwell won the bid to build a spacecraft to land on the moon. This was in response to the challenge posed by the US President John F. Kennedy to the American industry to send an American man to the moon and return him safely to earth. In 1965, they established a partnership with IBM to fulfil the requirement for an automated system to manage large bills of material for the construction of the spacecraft.

In 1966, 12 members of the IBM team, along with 10 members from American Rockwell and 3 members from Caterpillar Tractor, began to design and develop the system that was called Information Control System and Data Language/Interface (ICS/DL/I). During the design and development process, the IBM team was moved to Los Angeles and increased to 21 members. The IBM team completed and shipped the first release of ICS in 1967.

In 1969, ICS was renamed to Information Management System/360 (IMS/360) and became available to the IT world. The IMS database management system introduced the idea that application code should be separate from the data. The point of separation was the Data Language/Interface (DL/I). IMS controls the access and recovery of the data, but application programs can still access and navigate through the data by using the DL/I standard callable interface.

This separation established a new paradigm for application programming. The application code could now focus on the manipulation of data without the complications and overheads associated with the access and recovery of data. This paradigm virtually eliminated the need to store copies of the data. Multiple applications could access and update a single instance of data, thus providing current data for each application. Online access to data also became easier because the application code was separated from data control.

IMS DBMS became the bedrock of aviation MRO applications because it met the rigorous requirement, of performance and multiple access. The major aviation MRO applications used IMS, but later ADABAS also came in as a major player.

5.2.4 User interface

In the 1970s the internet was just about conceived by ARPANET and was not available for commercial or public use. However, IBM and others developed software and hardware which would allow communication between multiple users. One of them was part of the IMS suite of software and the other was called CICS (pronounced 'kicks').

CICS (Customer Information Control System) was designed and developed by IBM to meet the demand for online processing (Figure 5.1). It was generally made available in 1969. In 1972–73, CICS was supporting the large MVS-class customer. CICS was also an early supporter of the new 3270 Terminal technology. In 1970, CICS

Figure 5.1 A typical CICS login screen

announced its first support for COBOL and PL/1. This was the next best solution to internet, however limiting.

The airlines that were eager to develop an aviation MRO application had the means to satisfy their communications needs, which was critical to the MRO functionalities.

5.2.5 Communications network software

The communications network protocols were coming of age during the 1970s. In other words the architecture and semantics were being developed so that computers could talk to the terminals and other computers. Nowadays we take these for granted, but then TCP/IP was almost non-existent. Hence the organisations had to have their own dedicated communications network. The practice has not gone away though.

In the same decade, IBM announced the Virtual Telecommunications Access Method (VTAM). It is the software package that provides communications via telecommunication devices for mainframe environments and is the implementation of Systems Network Architecture (SNA) for mainframes.

SNA is IBM's proprietary networking architecture created in 1974. It is a complete protocol stack for interconnecting computers and their resources. SNA describes the protocol and is, in itself, not actually a program. The implementation of SNA takes the form of various communications packages, most notably VTAM.

VTAM provides an Application Programming Interface (API) for communications applications, and controls communications equipment such as communications adapters and communications controllers. In modern terminology, VTAM provides a communications stack and device drivers.

In effect, the necessary bits of the communications network software and protocols were available to the developers. These allowed the airlines to architect their MRO systems, which were accessible to the critical stakeholders: the engineers, technicians, supervisors, planners, etc.

5.2.6 Computer terminals

In 1972, IBM announced 3270 terminals. This was the beginning of online processing using video display terminals. Before this it was all hard copy terminals, such as the AT&T Teletype and the IBM 1050 or 2740/41. These terminals are euphemistically referred to as 'Green Screens' (Figure 5.2 shows one such terminal).

IBM stopped manufacturing these terminals many years ago, but the IBM 3270 protocol is still commonly used via terminal emulation to access some mainframe-based applications.

Figure 5.2 IBM 3270 terminal

The aviation MRO functionality was such that without these terminals any application would have been of no use. It was not like an accounting or inventory management system where batch processing would suffice. Hence the aviation MRO applications adapted these 3270 terminals as their user interface.

5.2.7 The computer

IBM announced System 360 (S/360) on 7 April 1964 (Figure 5.3). It was the front runner of mainframe systems. It was designed to cover the complete range of applications, from small to large, both commercial and scientific. The design made a clear distinction between architecture and

Figure 5.3 IBM S/360

implementation, allowing IBM to release a suite of compatible designs at different prices. Then S/370 arrived (Figure 5.4).

IBM S/370 was announced on 30 June 1970. It was an improvement on S/360 with better and faster processing and memory management. It maintained the backward compatibility and added: dual processor capability, a large main memory using integrated circuits instead of magnetic cores, support for virtual memory and 128-bit floating point arithmetic.

IBM S/370 could do the ground work that an aviation MRO application required.

5.2.8 The aviation MRO applications

In the 1970s the building blocks for aviation MRO application were ready and available. However, they were not cheap. The best part was that the functional specifications

Figure 5.4 IBM S/370

for aviation was also ready and there were experienced aviation engineers who were keen to help build these applications.

There were several initiatives to build the applications, but the noteworthy are: SCEPTRE, MEMIS and MERLIN. The others did the job but remained confined to their own organisation. These three proliferated and were sold to other airlines and organisations.

SCEPTRE and MEMIS still survive in some form or other but MERLIN graduated to MAXI-MERLIN and took one step forward in technology. It replaced IMS with ADABAS for database management system. Actually it was the ADABAS version, i.e. MAXI-MERLIN, which was sold to more than twenty airlines.

All the aviation MRO applications were designed and developed in-house. They were bespoke. However, they used the same building blocks and to some extent look and feel identical. The similarities were not just superficial. Their data models and transactions were also similar. This was because they used the same specifications, as the aviation MRO processes were highly standardised by then (we discussed this in the earlier chapters). Now they use the same building blocks to build the applications. The standardisation just happened but unfortunately it did not happen to the level where they could be compatible with each other.

One must keep in mind that MSG-3 was already available to the industry. This was way ahead of MRP-III. In other words, the aviation MRO processes were standardised before the advent of business process standardisation in any other industry. It seemed like the business requirement specifications were waiting for the technology to come up to speed, which it did in the 1970s.

Before COBOL arrived on the scene the coding for aviation MRO functionality was near impossible. FORTRAN and ASSEMBLER could not have coped with the requirements. Similarly, without a proper database system an aviation MRO data could not be managed using the filing systems, especially when complexity and performance are the major concerns. Also, it required a computer like S/370 with high processing power, communication protocols like SNA and software like VTAM to be able to reach out to all the user community; and an operating system like MVS to be able to orchestrate and execute this myriad of software.

The aviation MRO was riding on the crest of the IT at that time. It was a great convergence. And the major airlines opened their purse strings and launched into application development, using the cutting edge IT of the time. Let us discuss some of the major aviation MRO applications that were developed during the decade of the 1970s and earlier part of the 1980s. These were:

- SCEPTRE (System Computerised for Efficiency, Performance, Tracking and Recording of Engineering data) developed by Republic Airlines in 1973.
- MEMIS (Materials, Engineering and Maintenance Information System) developed by Alitalia in the early 1970s.
- MERLIN developed by USAir in the early 1970s.

There were others like MAXIS by British Airways, EMPACS by Cathay Pacific, etc. However, their impact on the aviation MRO industry was minimal compared to the above three. Nevertheless, we will discuss them as we further elaborate the era of the bespoke applications.

SCEPTRE and its 'perfect' world

SCEPTRE was designed around the concept of a 'perfect' aircraft, hence the core functionality was configuration management while the rest of the modules provided the supporting functions. It took advantage of the fact that ATA had clearly identified an aircraft's configuration using zones, chapters and sub chapters (please refer to ATA's iSPEC 2200 for details). So for each aircraft type the system would allow setting up a perfect model, i.e. with everything going right the model represents how the aircraft should be. Then the actual part numbers and serial numbers were fitted into the system the same way they were fitted in an actual aircraft recognised by its tail number.

It was an ideal system from configuration management perspective and until today nobody has come up with anything better. However, like any ideal solution it required a significant amount of effort and skill to set up a 'perfect' aircraft. In one instance, in the early 1980s it would have cost approximately US$100 000 per aircraft type just to set up the 'perfect'.

SCEPTRE was engineering focused and did not do great justice to the other aspects of the aviation MRO requirements. However, it was one of the best aviation MRO applications ever created. Even today very few application come anywhere close to its perfection.

Even though the objective of building SCEPTRE was to support Republic Airlines, it was realised that it could fill the gap in the aviation MRO market, where there was nobody selling to other airlines. It acquired a Trade Mark in 1973 (though it is abandoned now) and was packaged and launched for sale. It acquired more than a couple of dozen customers in a very short time.

The technology used to build SCEPTRE was COBOL for programming, IMS for database and it ran on MVS, using all the other necessary communications software and hardware. It was a big mainframe application. Therefore, unless the airline was large enough it could not justify the investment required. Every piece of software and hardware was top-of-the-line and expensive.

In the early 1980s smaller upcoming airlines like Gulf Air wanted to buy SCEPTRE but could not afford the large footprint. They demanded a smaller footprint. Republic Airlines then responded by hiring five software professionals from Pan Am and put them to work in sunny Florida to create a smaller version of SCEPTRE, which would use VM/VSE instead of MVS and ADABAS instead of IMS. Gulf Air from Bahrain was the main investor because it wanted the application badly as it was in the process of setting up an aviation MRO organisation in Abu Dhabi. The project was never finished. The airlines were trying to develop a 'product', which was not their core competency, and predictably it failed.

Later Republic Airlines was bought by Northwest Airlines and Northwest decided to pull SCEPTRE out of the market.

They also withdrew support to customers, who were left to fend for themselves. Northwest later went on to re-engineer the application with IBM's support but only for in-house use.

SCEPTRE was leading the application until the early 1980s. But then it could not keep up with the pace of progress in IT, and it went into what is known as the 'legacy' world.

MEMIS the workhorse

Not much is known about the antecedents of MEMIS and it did not have anything esoteric either. It was an offshoot of a material management system, which conformed to the aviation MRO functionality requirements. Sometimes it felt that it surreptitiously came into being riding the technology wave.

Like SCEPTRE, it used COBOL, IMS and S/370 as its technology platform. MEMIS was packaged and sold to more than a dozen airlines and some of them still survive. There never was an initiative to re-engineer the application because Alitalia did not pursue this non-core business of selling software.

MERLIN and its 'magic'

USAir decided to upgrade its material management system and bring it up to speed with the technology and aviation MRO functionality. MERLIN was born. It used PL/1 for programming and IMS for database on an S/370 platform. This application was sold to less than half a dozen of the airlines.

USAir, realising the gap in the market, decided to exploit the situation. They decided to improve on the package by replacing IMS with ADABAS and PL/1 with a third generation language, NATURAL, and rebranded it as MAXI-MERLIN. This packaging was right and MAXI-MERLIN was the hottest selling aviation MRO application of its time, at one million dollars a time. The airlines would buy the package including the source code and training. The business model was that once the airline bought the software, they were on their own. MAXI-MERLIN was sold to more than forty customers, a record.

The US Air Product & Services organisation was established in 1980 when Richard Abruscato joined US Airways (US Air at the time). The initial offering (1980) to the industry was an IMS version of US Air's M & E system which was licensed by Ansett Airlines located in Australia and Flying Tiger located in the USA. After realising that that operators and third-party maintenance providers would be looking for platform portability, a platform that would correlate with an operator's IT strategy, MAXI-MERLIN was redeveloped utilising Software AG's Natural which was a four-generation coding language and utilising Software AG's database Natural, a multi-functional database which essentially created an integrated database with relational capabilities which were far superior to other databases on the market at the time the solution was re-written. Software AG's Natural enabled programmers to be up to 14 times more productive than the industry standard at the time, which meant a faster and higher quality of system development and modification. Richard recruited two US Air employees to join US Air's Products & Services organisation to lead the redesign of MAXI-MERLIN. Mr. Jim French, the US Air Materials Expert and a US Air Maintenance Expert, joined the group in 1983 and took responsibility for the redesign, which was completed approximately fourteen months later.

The re-written MAXI-MERLIN was made up of seven integrated functional modules which allows for integration of all material management, purchasing, maintenance and engineering functions. The following are the seven functional modules which made-up MAXI-MERLIN:

- Material Service Control Module
- Component Control Module
- Modification Control Module
- Maintenance Activity Communication and History Module
- Shop Planning Module
- Work Cards Module
- General Purpose Module.

If a customer needs a corporate financial package, US Airways would deliver MAXI-MERLIN Plus which included the above modules, plus a General Ledger Module, Accounts Payable Module, Accounts Receivables Module, Budget Module, and a Fixed Assets Module. If a Production Planning and Control functionality was also required, US Airways would deliver Ultra Merlin, which included all MAXI-MERLIN Modules, plus Parts/Service Mater Data Module, Bill of Maintenance Module, Work Centre/Routing Module, Rough Cut Production Schedule Module (long range planning), Material Requirements Planning Module, Customer Maintenance Order Management Module, Maintenance Order Management Module (medium range planning), Maintenance Scheduling and Control Module (day to day planning), Maintenance Order Costing Module, and a Batch/Lot Control Module. The first MAXI-MERLIN customers to select and implement the redesigned MAXI-MERLIN were Kuwait Airways and BWIA.

Over the years the MAXI-MERLIN group increased to twelve employees. This group, me included, were brought on board to increase business functionality and to continually enhance and add functionality based upon requirements generated from customers, potential customers, regulatory agencies, manufactures, etc.

US Airways gave the customer the ability to select what functionality was required to meet the customers critical business requirements. US Airways invented and offered to the aviation industry, the industry's first M&E/MRO ERP Solution.

During the height of the Merlin aviation market availability, more than 40 diverse customers (airlines and third-party maintenance providers) utilised the solution. The airlines include FedEx, Southwest (who just recently announced that they will be implementing TRAX), AeroMexico, US Airways Regional Carriers, EgyptAir, Kuwait Airways, LOT Polish Airlines and Royal Air Maroc, who still utilise MAXI to manage and control their fleet of aircraft.

When US Airways outsourced all IT to SABRE in 1997, the MAXI group, as well as the solution became the property of Sabre. Over time all members of the original MAXI group left and after a number of years the support of MAXI and MAXI customers were no longer offered by SABRE.

Later, MAXI-MERLIN was acquired by SABRE, but currently it is not sold actively.

The others and their impact

There were other airlines like Cathay Pacific and British Airways who followed the technology trend of the time and created function-rich aviation MRO applications. However, they were not very successful in selling it to other airlines.

British Airways did not manage to sell its software but Cathay Pacific did manage to sell its EMPACS (Engineering and Maintenance Planning and Control System) to around half a dozen customers, Emirates being one of them.

EMPACS used PL/1 for programming, IMS for database and S/370 as the platform. However, it moved quickly to ADABAS, which enabled it to sell to other airlines.

Cathay, as was the custom, sold the entire package with source code and did not provide any support. It was after all a non-core function for them.

There was an attempt to re-engineer and repackage the software in partnership with SABRE, but this initiative never left the drawing board. A blueprint was created but later abandoned. Cathay then abandoned EMPACS in the late nineties.

5.2.9 Conclusions

There is an obvious trend in the application development carried out by the airlines. First of all, there were no independent vendors and the airlines have had to fend for themselves, which was not unusual at that time. Secondly, the airline, especially the larger ones had the capacity to invest in the best-of-breed technologies. And, finally they had a standard process, which could be coded in a relatively short amount of time.

The airlines unwittingly became vendors for aviation MRO software in the 1970s. However, they abandoned the scene once the IT moved on and the business imperatives changed with the de-regulation. The great IT journey lost its two major strengths: surplus money and IT skills (in some cases the entire IT department was outsourced).

However, the airlines created high performance and function-rich aviation MRO software, which the current software vendors are finding difficult to match.

5.3 The vacuum and the minnows

IT moved from the mainframe to the client/server. The relational databases came in and PCs, with user friendly interfaces, started replacing the IBM 3270 terminals. But nobody had an application for aviation MRO that would exploit these technologies. The airlines were desperately looking for solutions for more than a decade starting from the early 1990s but nothing worthwhile was available, and even when there were some they could not replace the legacy applications.

The option was to re-engineer the legacy systems or buy into another initiative by a software vendor. Moreover, the airlines shied away from making their own investment because the de-regulation had struck and they did not have the necessary buffer to invest in non-core risky ventures. The airlines were withdrawing from being software vendors and no major vendor was stepping in to fill this gap. None of the big software vendors seemed to be interested in the aviation MRO market. This created a vacuum.

The vacuum was not there for the faint hearted though. There was a skewed development which had taken place and its result was the major inhibitor.

5.3.1 The technology skews

Once the SCEPTREs, the MAXI-MERLINs and MEMIS were sold and deployed they took on a life of their own, since each customer had the source code and a team of IT professionals to support and operate the application. In technology terms, the time froze as there were hardly any upgrades from the vendors. In this case the vendors were the airlines that created these wonderful applications and who had quietly left the scene abandoning their responsibilities under various circumstances.

On the other hand, each of the customers kept adding functionalities over the years. Some of these enhancements were the same for all the airlines but were duplicated because each had to fend for themselves as far as the aviation MRO software was concerned.

The result of these activities was that while the technology remained stuck in time, the software were getting more and more rich in business functionality. This was a sure shot method of claiming a grand legacy status.

It is interesting to note that though they did not keep pace in the area of programming language, database and operating systems, they kept improving the performance of the application, either by increasing the memory or disk or the processing power. In some cases they improved performance by optimising the codes.

The net result was that each airline and MRO organisation had a formidable legacy aviation MRO application. They were functionally rich, had excellent performance and throughput and the workforce was conversant and comfortable using them. They became king, and could only be replaced if they died.

This was ominous for these legacy applications. The death knell was sounded by the IT industry in the 1990s. This was with the advent of new programming languages, relational databases, UNIX, PCs and client/server technology.

5.3.2 The state of limbo

While IT was forging ahead at breakneck speed, the aviation MRO application world went into limbo.

The airlines were neither equipped for nor inclined to any kind of investment in research and development on this front. It was recognised that IT was not their core competence to start with.

The existing aviation MRO systems were formidable competition to any vendor-created software. A new vendor with any software could not match either in functionality or performance or both. And then the airline management did not see any benefit in replacing the application when they were told that they will have less functionality or performance; even if it was miniscule. The technology aspect was appreciated by the management and for a good reason, it did not make financial sense.

A serious software vendor was up against three immovable barriers: (1) he could not match the depth of functionality of a mature legacy application; (2) The client/server technology could not match the performance of the mainframe; and (3) the engineers and technicians had become experts in the use of the green screen based systems and were not willing to change (one of the guys told me that he hated the mouse and the colourful screens). They were so used to the application that they would type on without even looking at the screen. So the software vendors, including the large ones sidestepped and avoided this market.

After that, the rot started to set in. Sensing this, the opportunists of the software world started biting off small chunks of the integrated business process of aviation MROs. Most of the organisations ended up running more than three hundred software systems to keep their business going.

Surprisingly though, even the threat of Year 2000 (Y2K) did not make a lot of impact on the legacy systems, and they survived.

5.3.3 Some pioneering attempts to leave the limbo behind

There were several genuine attempts in the IT industry to break the limbo, which, for one reason or another, did not succeed. The notable amongst them are: United Airlines and Qantas joining hands to take up the challenge; Sabre Group's ambitious M&E club; an attempt by a South African software company to bring MAXI-MERLIN to the client/Server platform; and an initiative by North West Airline, in collaboration with IBM, to re-platform SCEPTRE. They were all well-funded and pragmatic attempts, but they required a long period of development. The business customers might have waited, the information technology, however, did not. Most of the technologies, which were the core of these initiatives, became obsolete even before they were implemented.

EMSYS; short for Engineering & Maintenance System, was the name chosen by the joint team from United Airlines and Qantas, in the late nineties. The idea was elegant as well as revolutionary from the industry perspective. The idea was to build the application from scratch using the latest technologies; in other words client/server and related technologies. Internet and three tier technology was still just a glint in the eyes of the IT industry. In order to successfully deliver the system, a separate entity called Qantec was created. Qantec was tasked to build the system. It comprised of software engineers from Qantas and United Airlines. Since it was based in Australia, it was mostly Qantas driven.

The core of EMSYS was the master data structure built around ATA's codification of an aircraft's position, instead of part numbers. Since AMTOSS, the guiding standard for Task Cards, is also specified around this codification, the Technical

Document part of the system came up beautifully. However, it became too cumbersome and complex for implementing Engineering Planning and Inventory Management functionalities.

Because of these difficulties the system was never completed. The rights to the Technical Documentation module were bought by United Airline as the joint venture collapsed. Qantec supported Qantas for a time, but in the end none of the results of this venture reached the Aviation MRO market.

In a parallel move, another initiative was begun by the Sabre Group. Sabre had joined forces with Cathay Pacific and was promising a new platform for M&E application to other airlines, including Emirates, to form an M&E club. The entry price for membership was one million dollars, which was a lot for an airline in the late nineties. There were almost no takers, but Emirates joined in as an observer. There were workshops held in Hong Kong and Dallas.

After all these activities, the only outcome was a blueprint, which is rumoured to be used as a basis for CMRO. The M&E club never fully formed. Cathay went on to acquire a COTS application. Sabre gained MAXI-MERLIN as part of its acquisition of USAIR's IT division, which completely changed its focus. There was no need to develop any application from scratch when an already built application was under their belt.

Then a South African IT company that had been associated with Software AG, the owner of ADABAS (the database engine that MAXI-MERLIN used) took up the challenge of moving MAXI-MERLIN to the client/server platform. They were successful from the technology point of view but did not go far in the market. They could only replace some of the existing MAXI-MERLIN installations and soon dropped into the abyss of legacy applications.

North West withdrew SCEPTRE from the market but initiated a project with IBM to re-platform the application. Not much is known about this initiative and it did not generate much interest, mainly because it was reserved only for North West's internal use.

There were other attempts as well, but they were never publicized and the market took no interest in them. So the vacuum continued.

5.3.4 The swarming of the minnows

In the meantime, that is from the late 1990s, the outlook towards IT was changing and the drive for outsourcing the non-core activities was intense. The aviation MRO organisations were being created based on the same principles: maintenance of aircraft is not an airline's core competence, it needs to be outsourced. The airlines then hired off their engineering departments and created separate entities. This had an added advantage that these entities were free to take on new jobs from other airlines. So their business increased and so did the need for IT. However, they were not so rich as to be able to create bespoke software and maintain a large IT department. The quandary was that most of them had to depend on the legacy systems, which required support. There was a shortage of PL/1 and COBOL programmers and an absolute scarcity of IMS experts. The MRO organisations could not afford to keep a large IT, with expensive staff, hence they also resorted to outsourcing.

At the same time the hunt began for COTS (Commercial-Off-The-Shelf) software for aviation MROs, which was non-existent. As it is said, nature abhors a vacuum. So suddenly the so-called COTS software started appearing at an unprecedented rate. The demand for COTS software also encouraged a significant number of Aviation Engineers to quit their jobs and start software companies. The good thing was that a number of good software came about. However, they did not quench the thirst of the aviation MRO industry.

Today, there are more than two hundred aviation MRO COTS applications in the market. Some are extremely focused and niche but only a few can claim to cover the entire gamut of aviation MRO functionality. They still fail to match up to the legacy applications built in the 1970s. As pointed out earlier, the legacy system can only survive for so long, so many MRO organisations have opted for whatever is available in the COTS market. I will refrain from talking about specific software for privacy reasons; however, I will further elaborate on generic terms.

Since the big boys, SAP, ORACLE and IBM were still on the sidelines, the proliferation of aviation MRO software was stupendous. However, since each one was small, only a few customers (some have only one), it would not be enough for their survival.

So when the frustrated aviation engineers took up programming, hundreds of MRO software companies burst onto the scene. They all started on client/server technology, which never did surpass mainframe performance. But then after the Internet arrived along with three-tier architecture, they jumped on the band wagon, which helped them improve their performance.

These multifarious COTS applications were called best-of-breed because they designed to perform specific tasks for aircraft maintenance. They mostly did not cover the support functions required for aviation MROs hence needed to be integrated with other applications, including ERP software.

5.4 The big boys get interested

After the tsunami of ERP implementations across the world subsided, SAP and ORACLE started paying some attention to the aviation MRO market as well. This was after the turn of the century and after the scare of Y2K. By then Y2K was behind us and most of the major organisations had gone the ERP way.

Notably, there were four major players who singly or jointly started probing into the aviation MRO market and looking for a leadership position. These were IBM, SAP, ORACLE, and Boeing. IBM initially just wanted to be a systems integrator but later found that it had a product, MAXIMO, which could be positioned for this market. SAP and ORACLE were product focused and Boeing wanted to be a facilitator.

5.4.1 Boeing gets going

Boeing felt that there was a cavernous gap in the aviation MRO market – rightly so – and decided to pitch in. In the late 1990s, Boeing went into partnership with an Indian company called RAMCO and decided to build an aviation MRO application from scratch. However, after the 9/11 disaster they discontinued the project. There could have been other reasons too; RAMCO says it was the downturn in aviation interest after the Twin Towers were hit. They kept the project going on their own and ended up with the RAMCO Aviation software, which is being marketed worldwide.

But Boeing had not given up on the aviation MRO as yet and had only changed tack. Around 2002 it acquired aviation MRO software called AeroInfo Systems and established a partnership with MXI Technologies of Canada. The rational seems to be to build on existing software, which had the right technology platform, rather than build from scratch; hence the idea to build on AeroInfo with MXI. Since Boeing did not have the will or inclination to invest in development staff, it chose MXI to carry out the product development. This product was named Enterprise One. IBM Global Services was to be the implementer and systems integrator of choice. However, this venture also did not come to fruition, so MXI Technology went ahead and built an application called Maintenix. Hence Boeing spawned two software products: MXI-Maintenix and RAMCO Aviation. Both of these are now prominent players in the aviation MRO COTS software market.

5.4.2 SAP picks up the scent

It was now SAP AG's turn. They decided to create an industry specific module called A&D (Aerospace and Defence) and go after the aviation MRO market. It was almost parallel to Boeing and ORACLE's efforts who were also trying to get a piece of the pie.

SAP by then had a large customer base and many among them were airlines and MRO organisations. It then roped in a few big MRO organisations like Lufthansa Technik, SIA Engineering Co, British Airways, SR Technik and JAL to use the SAP platform to deliver aviation MRO functionality. Since SAP does not believe in implementing its own software, it needed to find a systems integrator. It was Pricewaterhouse-Coopers (PwC), later acquired by IBM, which was selected as the system integrator of choice for implementing the SAP-based MRO applications in almost all the sites.

Each of these implementations was very large, complex and expensive (the unofficial estimate is that they cost tens of millions of US dollars each).

SAP germinated in IBM as a 'Special Application Project', and is based on the processes defined by MRP-III

(Manufacturing Resource Planning), which essentially takes the premise that all the best practice processes required by any business organisation are identified within SAP and they just need to be configured and enabled to deploy an IT system. In other words, the organisation had to change its current processes to map with SAP or the best practice processes. This was not an issue with most of the business organisations, including OEMs like aircraft and engine manufacturers. However, this philosophy came unstuck with aviation MRO.

Aviation MRO processes are based on MSG-3 and there was no way an aviation MRO organisation could change its processes to conform to SAP-based processes, which are MRP-III, even if it wanted to. Any change in the specified process could cost them their certification as well the ability to meet the criteria for airworthiness.

Therefore SAP had to change, but this was not easy. SAP came up with their A&D module, later called SAP MRO and eventually partnered with AXON (now HCL-AXON) and included their product iMRO as the MRO enabling module.

5.4.3 ORACLE sees the future

ORACLE's early foray into the aviation MRO market was not at all successful as the initial offerings were not well received by the market. However ORACLE, after a gap of two to three years, decided to change its strategy and came to the market with an entirely new offering: CMRO (Complex MRO). This was a complete redesign.

ORACLE moved away from ERP-based design to create CMRO as an almost independent best-of-breed module. In theory it can work with SAP as well. However, it relies on the foundational application layer of ORACLE Application for system integration with other modules.

In doing so, ORACLE skirted the problem of MSG-3 vs MRP-3 and provided an offering specially targeted at the aviation MRO market, and has been strongly pushing in.

5.4.4 IBM's accidental entry

IBM's proclaimed strategy of not owning applications continues; however, when it acquired MRO and its software MAXIMO, it found itself playing in the fringes of the aviation MRO market. MAXIMO was bought by IBM as a replacement for Peregrine, the software for supporting help desks. This was because IBM was using Peregrine extensively to support its help desk service provisioning and Peregrine was acquired by HP, so IBM needed a replacement. IBM looked around, found MAXIMO and bought it.

The fact is that MAXIMO does not just do service management, it also supports maintenance. And just before its acquisition it had acquired the configuration management software called Raptor. This was with the intent of entering the aviation MRO market. MAXIMO was already being used by Roll Royce and it was reasonable to assume that it would be possible to gain a foothold in this new market, where all kinds of players were storming in. It did have some success but could not make any deep impression on the market. However, IBM has yet to formulate a defining move based on a clear strategy and it is not clear whether IBM is fully committed in this market.

5.5 The active vendors

Out of the swarm of winnows, some rose faster than others and established themselves in the rarefied aviation MRO application software market. Earlier in this chapter we mentioned RAMCO and MXI. These vendors, along with TRAX and AMOS, made some serious investment in developing best-of-breed MRO applications.

5.5.1 AMOS morphs out of SWISS IT

Like any other airline in the 1980s, Crossair had IT which not only managed the applications and hardware but also created products. This happened when Crossair's IT jumped on the client/server band wagon and created aviation MRO software, which was aimed at meeting their internal requirements. However, it soon realised that there is a gap in the market and started venturing out and selling this software to other airlines and MRO organisations. Relatively they were very successful. Even today, AMOS has been sold to almost ninety customers; a large number for the aviation MRO market.

The history of AMOS began in 1989 in Crossair's IT department, where its foundations were laid. Crossair later took the name SWISS: Swiss International Air Lines Ltd. In 2004 SWISS decided to hive off its IT department and created a 100% owned subsidiary called Swiss Aviation Software Ltd (Swiss-AS). Today it operates independently and with a staff strength of around sixty. It has partnered with strong MRO organisations like Singapore-based STA and other software service vendors for implementation.

Though still a client/server aficionado, AMOS has a strong presence in the aviation MRO market as one of the best-of-breed applications.

5.5.2 MXI moves on with Maintenix

In 1996 a small Canadian software company was formed in Ottawa, which won a contract to create maintenance software for the Canadian Air Force. This was called

MXI Technologies. The software they developed was called Maintenix, using the latest available technologies at the time.

MXI has established itself as a major aviation MRO player by partnering with Boeing and IBM in the initial period. Some of the large airlines like LAN-Chile and Qantas have taken to Maintenix to support their MRO operations. MXI has been profitable for a significant time since its inception.

Later, however, the partnerships with IBM and Boeing fell through, though Boeing has declared intentions of using Maintenix for its support services.

5.5.3 RAMCO decides to diversify

RAMCO is a well known brand in India, though not for software or aviation MRO. It was known as a major cement manufacturer from the south of India. Around the early 1990s RAMCO decided to get into software development and came up with a low cost ERP application, which was very successful in India. Later they got interested in aviation MRO and partnered with Boeing to build a best-of-breed software for aviation MRO. However, the partnership broke down in 2001. The reason assigned is the down turn in aviation industry due to the 9/11 disaster. Boeing seemed to have lost its appetite for developing MRO software from scratch.

However, RAMCO did not stop its development of the software and went ahead without Boeing. It created RAMCO Aviation software, which managed to capture a fair share of the market consisting of middle sized airlines and MRO organisations. One of its major wins is Air India.

RAMCO, like MXI, uses the latest technologies and significantly rich functionalities. However, it has not been able to establish itself with larger organisations.

5.5.4 TRAX on track to capture market

TRAX came in riding the client/server wave, backed by a group of aviation MRO engineers. It is best-of-breed software with significant coverage of engineering functionalities. The software had a good traction in the aviation MRO market. TRAX was one of the first to move away from mainframe and adopt client/server technology. From an engineering perspective, the software supported almost all the functionalities. Like any best-of-breed software, it depends on other applications to provide non-engineering functions.

TRAX proliferated and captured a significant share of the market consisting of middle and small airlines and MRO organisations. However, it could not make a dent in the larger organisations.

Client/server technology, one of the main reasons for TRAX's success, later became a hindrance and they struggled to move into the three-tier web-based technology.

Currently, TRAX has a good presence in the aviation MRO market and is pushing ahead aggressively.

5.5.5 VisAer sparkles then vanishes

VisAer was also formed in the early 1990s and went into the aviation MRO market with a great ambition, like MXI and RAMCO. Its platform was developed using the latest technologies available at that time and its close association with Jet Blue helped them develop the MRO software. Most of the MAXI-MERLIN team joined in the effort after MAXI-MERLIN became one of the software offerings from SABRE, and Robert Abruscato died. The MAXI-MERLIN team was unable to fully integrate into SABRE and was effectively leaderless. It seems the team members found solace at VisAer.

173

VisAer managed to win a few prestigious customers, notably China Southern and Qantas along with some medium and smaller airlines. However, it could not grow much further and was bought in 2008 by an Indian software developer and rebranded.

5.6 Summary

In this chapter we talked about how the aviation MRO software applications came into being. They were created by airlines driven by the need to support their maintenance and engineering activities. These software developments were mostly bespoke but were also early adopters of the latest IT of the time. Initially they were hugely successful, but later became victims of their own success as IT moved on.

Then there was the impact of de-regulation of the commercial airlines, which ended up in severe competition, which in turn brought the margins and cash flow down to a bare minimum. The airlines, along with others, listened to the strategy gurus and got influenced by the mantras of 'core competencies' and 'outsourcing'. Thus any further investment in software development, especially for MRO, died out. Everyone started looking for COTS software, which did not exist.

Identifying this demand a plethora of entrepreneurs launched themselves into the aviation MRO market. Many lost their way and foundered but some survived, although there was and is no leader of the pack. The two majors, SAP and ORACLE, also came into the fray, a little late though. But they still have not been able to establish a clear leadership or have not been able to garner a substantial market share.

Then we discussed some of the major players. It seems most of them were spawned by the OEMs and the airlines themselves. The specifications were provided by the OEMs like Boeing and the airlines, but instead of developing the software themselves, they engaged other software companies to independently develop for them, thus reducing their risk and not getting involved in the marketing and sales side of the product. However, for all practical purposes the scenarios were similar to in-house development, but by outsourcing it to external developers who would also invest in the development and take the risk. The OEMs limited themselves to just providing the specifications.

This chapter tried to answer an oft repeated question: 'Why is it so difficult to find the right software application for aviation MRO?'

5.7 Note

1. Kena Upanishad, *Upanisads*, Oxford University Press; translated from Sanskrit by Patrick Olivelle; 1996.

6

The current aviation MRO IT landscape

Abstract: This chapter highlights the available and in-use IT solutions for aviation MRO. The chapter provides an overview of current landscape; describes issues and challenges for system architecture and integration; highlights commonly used technologies and services; identifies enablers to significantly improve the use of software in the maintenance of aircraft; and describes some of the promising technologies for aviation MRO.

Key words: Enterprise Resource Planning, best-of-breed, legacy.

It moves—yet it does not move. It's far away—yet it is near at hand! It is within this whole world—yet it's also outside the whole world.[1]

In the last chapter, we covered the history of aviation MRO software applications. In this chapter we will discuss the applications from a solutions point of view, in order to bring out the nuances and difficulties that the current MRO IT landscape has.

A solution, as we understand, is an IT solution which enables and automates the aviation MRO functions and processes. This consists of software, data, process flow and the technology infrastructure. Interplay of all these results in automation.

As noted in the earlier chapters, the aviation MRO software applications can be categorised as either being legacy, best-of-breed, or integrated ERP-based (Enterprise Resource Planning) solution. These three types of solution have their own peculiarities and issues, which we will discuss shortly, and all need technology platforms such as servers, network, operating systems middleware, interface devices, etc. An entire architecture which can deliver all that is needed to keep an aircraft airworthy, effectively and efficiently.

We discuss the three types in the following sections.

6.1 The legacy solutions

A legacy is:

> something such as a tradition or problem that exists as a result of something that happened in the past or something that someone has achieved that continues to exists after they stop working or die.[2]

There are a significant number of aviation MRO sites, where the 'legacy' applications are still thriving. As mentioned earlier, the heyday of the MRO or M&E software was the late 1970s, when the money allocated for M&E was bountiful, the processes had been standardised, and innovation was in the air. These applications were so robust that even now, after three decades of their being developed, it takes huge amount of money and guts to try and replace them. These applications were built by the real artisans, who had a complete view of the requirements and knew the strengths and limitations of the technology of the time.

These artisans do not exist anymore. They were a product of the time and situation, which does not exist anymore, so there is no chance that such artisans will be born again. Adding to that, most of the technologies behind these solutions are obsolete, hence they are destined to die. And they are dying. If we look around we witness the fate of the SCEPTRE, MAXI-MERLIN, MEMIS, EMPACS, and the likes. They are all on there way to obscurity, if not already consigned to their graves.

One of the aspects of these legacy applications that should not be ignored is the fact that today, where these applications exist, they do not look anything like they did when they were conceived and born. Hundreds of thousands of person days have been spent over the decades to constantly add and modify their functionalities, and now they are robust and reliable workhorses. However, they are blinkered and are unable to adapt to the new technologies and business models.

When these legacy applications were developed, the designers never even thought that an airline will NOT maintain its own aircraft. Since the FAA had mandated that it is the operator's responsibility to maintain the airworthiness of the aircraft it owns and operates, it was a given that an airline will always maintain its own aircraft. Hence, these applications were designed with the view that did not take into consideration the business model which is now prevalent: the airlines do not necessarily maintain their own aircraft and prefer to outsource it to the aviation MRO organisations. Most of the legacy MRO applications do not know how to manage this business model.

This became evident to me when we tried to implement one of the legacy applications, SCEPTRE, in GAMCO, an MRO organisation and a subsidiary of Gulf Air, Bahrain. GAMCO was modelled after the large and predominant MRO organisations such as HAECO and Lufthansa Technik and was the first ever aviation MRO venture in the Middle East. GAMCO later became ADAT and cut its umbilical cord with Gulf Air. It is situated about 40 km from the city of Abu Dhabi, at the periphery of the airport.

The first problem that struck us, the implementation team, was the problem of ownership. SCEPTRE did not have the facility to handle multiple ownerships. We toyed with the idea of implementing the functionality but it was too hard and expensive to implement. A major change like introducing the concept of ownership in an application which requires modifying hundreds of programs written in PL/1 is a gargantuan task and fraught with extreme risk. Notwithstanding this issue, SCEPTRE was ultimately abandoned by GAMCO and Gulf Air. It was replaced by MAXI-MERLIN. However, that decision did not fully resolve the problem of ownership but allowed working around it, which was easier to implement.

These legacy applications require a team of software engineers and programmers proficient in the legacy languages and techniques, such as COBOL, PL/1, NATURAL, JCL, ADABAS, IMS, VSAM, etc. These skills are now very hard to find. The number of people skilled in these areas is dwindling and no one even wants to learn these skills anymore. This not only causes concern in the area of system development and maintenance but also while migrating these applications to other systems. This skill set is needed to help migrate the data and map the necessary functionality with the new application. It is not just the skill that is required but also experience because none of these applications have robust documentation support.

These applications have survived for so long because they meet the users' needs and have wonderful performance and reliability records. They have survived longer than most of the other legacy applications because the core functionalities of the MRO processes have not radically changed over the decades. And then, the cost of migration or transformation is formidable.

There is one other aspect of these legacy applications worth noting. Over the years, they tend to morph into a nucleus, surrounded by hundreds of small applications. These applications could be as small as the macros in an MS Excel worksheet to maintenance planning sub-systems. Most of these sub-systems are created to augment the functionality of the core system, especially when it becomes too cumbersome or expensive to enhance the core application. The number of these satellite applications can be as many as three hundred or more and they do not function independently. They are tied to the nucleus, the core application, and in most cases have to be replaced if the core system is being replaced.

It is obvious that these legacy applications will have to be replaced in time. However, the challenge for the airlines and MRO organisations, who still operate these legacy applications, is how to replace them and with what.

6.2 Best-of-breed solutions

The phrase best-of-breed is:

> used for describing an animal that is judged the best example of its type in a competition or show or used for describing a computer product that is the best available of its type.[3]

In the context of aviation MRO applications, this phrase can only be used as a generic term to describe those applications which are not ERP solutions and focus on delivering MRO functionality. They are more of a pack of breeds than a specific breed, whether we look at them from the technology or functionality perspective.

On the technology continuum, some of these applications may also be considered legacy because client/server technology is now passé. However, they are different from the mainframe-based legacy applications. They are owned and supported by independent software vendors, which are also capable of keeping up with the technology. It is assumed that all these software vendors have either already adapted the web technology or they are going to shortly. Otherwise, those who do not will end up being replaced by some other application.

The common theme running through this pack of application software is that they focus on the engineering side of the MRO processes. Some of them extend into the area of material management. However, this is limited to bare minimum functionality, which essentially supports the engineering functions and not necessarily the full supply chain. They depend on other applications to provide the functionality, e.g. Human Resources, Finance, etc. Beyond that none of them have any similarity. They have different sets of transactions and each one of them implements the MRO process in its own way. Somehow, there is no standard approach that is visible in this pack of application software. Each one of them is unique.

Sometimes, it seems better to call this software offering 'point solutions' created by each vendor driven by entre-preneurial vigour or dragged in by circumstances. The point is that none of them have managed to comprehensively corner the market or satisfy the customers. There is hardly any aviation MRO organisation which is not going through an evaluation or health check process, unless they have invested heavily in an expensive ERP-based solution, which will be discussed in the next chapter.

One would assume that if an aviation MRO organisation has implemented a best-of-breed solution, that will be the end of their journey as far as software evaluation and search is concerned. However, there have been cases where these solutions have been replaced or the customer is not very satisfied and is looking for replacement. One of the cases in point is where LAN-Chile replaced VisAer with Maintenix. Both these applications belong to the same genre. Then Air Canada dropped Maintenix after trying to implement it for almost two years. It must be noted that Air Canada was one of the sponsors of MXI Technology, which is a Canadian company. AMOS is being constantly evaluated by their customers and in some cases replaced.

The trouble with the software vendors in this category is that all of them are very small, their staff strength is normally between 15 and 70 at the most. Their ability to invest is also limited by their size. So this puts a big constraint on them to win the market. They are also dependent on implementation partners and do not necessarily get into the application maintenance business. Therefore, their revenue is totally dependent on the licence sale and the annual maintenance fee. This maintenance fee is not very significant because their licence price is not very high. In other words, these companies are in the business of delivering a highly complex and large application, which an aviation MRO organisation needs – as evidenced in our previous discussions – but their ability to scale and provide support is very limited. The larger customers always have the question at the back of their mind: what happens if the vendor goes 'belly up'? And this is not just a rhetorical question.

One of the strategies that these companies adopt is to form a partnership with large established IT vendors like IBM to project stability and confidence in the market. But at the end they end up parting company or fall out for various reasons.

MXI worked very closely with IBM at Air Canada, LAN-Chile and Qantas but the partnership did not last. Similarly,

RAMCO worked with IBM at El Al and Air India but parted ways at different phases of the project. Somehow, this strategy has not worked. This could also be because IBM has not articulated its strategy in the aviation market very succinctly. IBM has taken a software agnostic view in this market. However, the smaller software vendors are suspicious of IBM, because it owns MAXIMO, which is potentially a competition to them.

Then there is an issue of integration with other applications. By definition, we know that the software of this genre only provides the core MRO functionality, hence it needs to integrate with other applications. This integration is not necessarily simple, especially if the platform used is client/server. Even with web-based applications the problem does not go away. System integration becomes one the critical success factors for implementation. Again, the size of these companies forces them to look for systems integrators or depend on the in-house IT capabilities of the customer.

6.3 Integrated ERP solutions

It was in 1990 that Gartner coined the term 'Enterprise Resource Planning' or ERP, which was an extension of MRP-III (Manufacturing Resource Planning), specified by APICS. The idea was that combining all of the core functions of a concern on a single enterprise-wide software suite enabled more effective management of a mid-sized to large enterprise. But this does not necessary mean aviation MRO organisations. The issue of MRP-III and MSG-3 has been discussed earlier. However, SAP and ORACLE have launched themselves in this market with their ERP-based offerings. SAP has partnered with HCL-AXON and ORACLE has gone all on its own to capture this market.

There are mainly two applications which can claim to be in this category: SAP MRO and ORACLE CMRO. RAMCO also claims to be in this category but it does not entirely fit into this scenario, or most likely not perceived as such because it positioned itself as best-of-breed software.

The basic premise of these solutions is that they are based on an integrated database. All the functions dip into the one single database to execute a transaction. Therefore, if one transaction changes the data, all the other transaction come to know of the change, without any additional interface requirements. Hence, the database is prebuilt and minimum customisation is required to deliver the functionality for each implementation, in theory. And, there is minimal need to interact with any other applications. The main message is: homogeneity and comprehensive functionality.

Having said that, in order to understand the implications of 'comprehensive functionality', we must be cognizant of the fact that there is no 'free lunch'. Up to now all the ERP-based implementations have been expensive: tens and in some cases hundreds of millions of dollars. There are three factors that contribute to this high cost: the basic licensing cost of the ERP software itself; the cost of customisation; and the cost of implementation. In spite of the software being all pervasive, some integration with other applications are also required. Let us discuss the two offerings separately.

SAP MRO consists of three application layers: the core SAP modules, the SAP MRO modules (previously SAP Aerospace & Defence) and iMRO (a product from HCL-AXON). However, Singapore Airlines, Japan Airlines, British Airways, Lufthansa Technik, SR Technik and some others have implemented SAP MRO, using only two layers, i.e. without the iMRO component.

While RAMCO tends to be more of a best-of-breed type of solution, sometimes it does project itself as an ERP solution as well. However, the aviation MRO market does not view this offering as an ERP-based solution.

Until the present, the cost of implementing ERP-based solution has been humongous and has taken years to implement. This also deters MRO organisations or airlines to adapt to this kind of solution. Even those who have gone on this path are not necessarily fully satisfied or have been able to fully realise the benefits envisaged.

6.4 The technologies

Almost all the aviation MRO solutions were initially developed on the mainframe and were fined tuned over the years. Hence, they were not only functionally rich but also their response times are very fast. Compared with legacy applications the response times and throughput using client/ server are much lower. This had been the bane of the new software, which tried to replace the mainframe-based applications.

Somehow even the tsunami of 2000 could not dislodge the mainframe applications from the aviation MRO market. The legacy applications ruled and survive even today, especially in the large organisations like American Airlines and Northwest Airlines.

However, the mainframe technology could only go so far. Hence the client/server technology later prevailed but was soon replaced by the web technology. Then over time the power of the servers kept improving and at least a reasonable level of response time was achieved.

In the area of operating systems, the transition from MVS to UNIX or MS Windows was as painful as moving from mainframe to client/servers. Not that they were independent of each other. However, UNIX did not match up to the robustness and reliability of MVS and became a major inhibitor to technology upgrade for the MRO applications.

In the field of IT there are three areas that have been constantly changing and they directly affect any application. aviation MRO applications were no exceptions. The three areas, briefly discussed earlier, are:

1. the hardware
2. the operating system
3. the network.

Currently, the aviation MRO applications straddle multiple generations of technologies. One can buy an application running on the mainframe as well as those using servers and even standalone PCs. Similarly, there are applications that run on MVS or Z/OS, UNIX and Windows. In other words, there are a myriad of applications using very diverse technologies to deliver aviation MRO functionalities.

There are applications such as TRAX and AMOS who have very high visibility and number of customers but still remain in the world of client/server technology. Their argument for this is that the throughput in client/server is better than the internet-based technologies.

There is, however, an attempt to play the catch up game. Some of the vendors are trying to adapt to eSignature and tablets at the user interface. But the core technologies are being attempted by very few vendors from among the best-of-breed. Surprisingly, unlike other industries, the best-of-breed software vendors are behind in technology to the integrated ERP-based software vendors like SAP and ORACLE.

6.5 So many solutions but no holy grail

In the field of aviation MRO application, even though there are multiple vendors and offerings, the clients (i.e. the aviation MRO organisations) are still not fully satisfied by any one or combination of the applications available.

The conundrum of best-of-breed v. integrated solutions applies to the aviation MRO industry as well.

It has been said that selecting enterprise computer systems is a bit like planning a vacation. Should you go for the 'packaged tour' with an integrated system from one vendor, or plan your own itinerary, the so-called 'best-of-breed' approach?

The best-of-breed option usually provides richer functionality, satisfying more users. But dollar savings, convenience, and efficient data sharing can make the integrated approach very appealing.

Integrated systems provide multiple applications with a common database and consistent user interface so that all modules have a familiar look and feel. The downside is that some applications may have anaemic functionality, causing users in these areas to become disgruntled.

Best-of-breed systems, designed specifically to excel in just one or a few applications, can also pose challenges, such as increased training and support, complex interfaces with other systems, duplicate data entry, and redundant data storage.

In the aviation MRO industry almost all applications can be termed as best-of-breed, which either try to extend their functionality by relying on the integrated system or the integrated system is extended to meet the complex requirements of the industry.

Generally best-of-breed solutions are more expensive than the integrated systems in other industries. However, in aviation MRO industry this is not the case.

Replacing an MRO application is daunting whether it is best-of-breed or integrated system. Even an upgrade costs a lot of money and time. The key issues for this are twofold: the need for data migration and harmonisation as well as the cultural change issues. The MRO data, because of interlinks, complex business rules and dependencies, poses a major hurdle in data migration, especially where the source and target data structure are even slightly different. And the quality of data in the incumbent system also adds to the problem. In order to successfully replace or upgrade an MRO application, significant effort is required to understand, cleanse and harmonise the incumbent data before migrating it to the target application. Experience shows that moving to an integrated system is generally a bigger task in terms of data migration than moving to a best-of-breed solution.

However, in a best-of-breed environment, skills to use and support multiple systems with different hardware platforms, operating systems, databases, and programming languages are in short supply compared to those on the integrated system.

The other aspect that affects this debate is the size and viability of the software vendors. A best-of-breed vendor is normally smaller than the integrated system vendor. The chances of a best-of-breed vendor going under are higher than the integrated system vendor. And where the best-of-breed systems are implemented, a number of other support systems have to be implemented as well, to create a comprehensive systems environment that can support the MRO functionalities. This translates to having to manage multiple vendors.

One of the key strengths of an integrated system is a common database. But this is not necessarily a differentiator anymore. Nowadays, the best-of-breed applications use standardised data structures and RDBMS like their integrated brethren and can provide efficient integration using SOA (Service Oriented Architecture).

While the strongest case for the best-of-breed option is richer functionality, integrated systems are starting to catch up with their 'boutique system' cousins.

In a nutshell, the current scenario is that the aviation MRO industry is struggling with not only the conundrum of the choice between best-of-breed and integrated systems, but also with the lack of a fully fledged function-rich solution in the market. There is no industry leader amongs the software vendors for this market and there are no distinct trends in the buying behaviour of these organisations on this front.

If we believe the claims made by the vendors, all of them are market leaders. However, from the perspective of market share there is no identifiable leadership in the aviation MRO applications market. Strangely enough there have been no mergers or buy outs in this market so far. Somehow, it feels that these applications play in the backwaters of IT.

The aviation MRO organisations are still waiting for someone to build a leading edge solution so that they do not have to go hunting all over the world for the right one.

6.6 Summary

In this chapter we discussed the current state of the IT enablement of the aviation MRO industry. The main points that come across are that the current trend is neither distinct nor discernible. As the stock market analysts will say, it is trending 'sideways', Hence the buyers find it difficult to decide on how to place their bets.

The IT industry still does not seem to have fully committed to the aviation MRO industry.

6.7 Notes

1. Isa Upanishad, *Upanisads*, Oxford University Press; translated from Sanskrit by Patrick Olivelle; 1996.
2. *http://www.macmillandictionary.com/dictionary/british/legacy*
3. *http://www.macmillandictionary.com/dictionary/american/best-of-breed*

7

Leveraging IT and shaping the future

Abstract: This chapter highlights the process model and best practice that can be used to support maintenance of aircraft. The chapter provides an overview of the future of IT enablement for aviation MRO; describes an ideal scenario and its challenges for implementation; highlights feasibility issues; identifies enablers to achieve the IT nirvana; and describes some of the promising futuristic technologies for aviation MRO.

Key words: architecture, MRO, APQC, Enterprise Data Model, Business Process Model.

The face of truth is covered with a golden dish. Open it, O Pūşan, for me, a man faithful to the truth. Open it, O Pūşan, for me to see.[1]

One of the ways that the *Oxford English Dictionary* defines the word 'leverage' is 'use (something) to maximum advantage'.[2] We use this definition in our context in this chapter, which is the theme of this book. This word can be substituted for any of the following: use, update, take advantage of, exploit, or apply.

In the previous chapters, we discussed the entire gamut of IT landscape in the aviation MRO industry. There we elaborated on the evolution of the aviation MRO applications amongst others, now it is time to talk about an ideal solution.

In the following sections we will describe the 'El Dorado' of IT enablement for aviation MRO. Whether this can be achieved or not is a different matter. However, it will surely help create a vision and a target to aim for. In describing the ideal solution, we have not pulled any punches and tried to keep the constraints at a minimum. This does not mean that it is based on ideas without bounds but the concept is built taking and understanding the essence of the IT enablement, using Enterprise Architectural principles, frameworks and methodologies. Some of the terms used in the following sections will be very familiar to IT architects but that will also be easily understood by others as well.

7.1 Airworthiness and information technology

The current thinking around airworthiness is compliance, compliance and compliance. This is a mindset that needs change. Airworthiness, in current market environment, is about compliance, efficiency and cost effectiveness. It is true that 'technically' an aircraft is airworthy if it complies with the safety standards, but the fact is that it is not 'worth' flying if the process of making it airworthy is neither efficient nor cost effective. Once an MRO organisation aligns its strategy towards the three aspects of airworthiness, it needs IT to achieve the goal.

The three proposed goals for airworthiness – compliance, efficiency and cost effectiveness – result in revenue growth, increased operating margin and efficient capital utilisation. These are the results that any organisation seeks and an aviation MRO organisation is no exception.

In this holistic connotation of airworthiness we can identify a set of metrics that would let us measure the performance of

194

an aviation MRO organisation. The main two metrics are: despatch reliability and aircraft utilisation. Let us discuss these two first because the other metrics, to be discussed later, will be somewhat supportive.

7.1.1 Despatch reliability

The ratio of the number of flights delayed because of technical faults to the total number of flights, expressed as a percentage. Delays caused by other reasons are not to be taken into account for this calculation.[3]

The objective of an aviation MRO organisation is to ensure a very high degree of despatch reliability. Nowadays, the aircraft operators demand 99%+ despatch reliability. In other words the operators do not want any delays which can be attributed to technical issues. It should be noted, however, that any delay which is less than 15 minutes is not counted. But soon, with the growth of number of aircraft and stiffer competition, this luxury might also vanish.

First of all the information related to the calculation of despatch reliability must be available and then, more importantly, all the information regarding the factors that contribute towards maintaining a high degree of despatch reliability must also be available at the right speed and right time. In this IT can and does play a significant role.

7.1.2 Aircraft utilisation

Normally for transport and commercial aircraft, it is the average number of hours that an aircraft is actually in flight during each 24-h period or per month. For combat aircraft, the reckoning period is a month instead of 24h (i.e., hours flown per aircraft per month). However, some air forces use the number of sorties instead of hours.[4]

The way the aviation industry is heading, very soon this metric will drive almost all the revenue that an aviation MRO organisation will earn. The operators want a simple approach so that they are relieved of their book keeping and convoluted contract negotiations. The message coming from them is: we will pay for the maintenance based on the number of hours it flies. In other words an aviation MRO organisation will earn more the more an aircraft is utilised.

Though deriving aircraft utilisation is simple, mapping it with the cost and revenue requires a massive amount of complex calculations to get it right. Here again, IT can come to the rescue.

7.1.3 The supporting metrics

Scaling down from the two main metrics, we see multiple metrics that provide visibility of the business operations. The list below shows an example:

- maintenance cost per flight hour
- service cost per flight hour
- material cost per flight hour.

Figure 7.1 shows how an aviation MRO organisation delivers value to its shareholders.

These metrics show how the business is performing, and when set against the industry benchmark they show how well the organisation is performing. Unfortunately, there is

Figure 7.1 MRO enterprise: delivering shareholder value

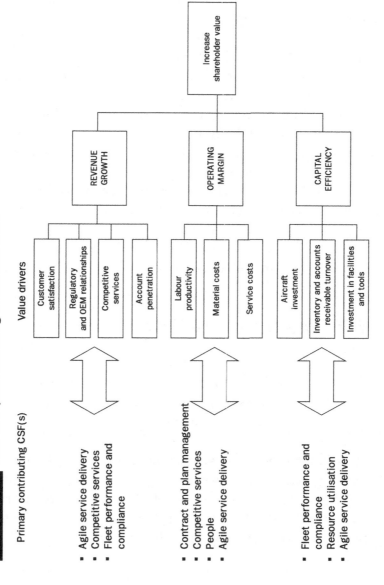

no universal benchmark available in the market. Some companies have developed some of the benchmarks but a comprehensive set of benchmarks which is universally agreed has yet to see the light of day.

This does not mean that IT cannot be leveraged, but it means that the beneficial affects of IT cannot be measured.

Having said that, let us talk about leveraging IT to ensure that an aviation MRO organisation provides an aircraft operator with maximum possible despatch reliability with optimum aircraft utilisation.

Before we jump into such discussions, let us keep in mind that the organisational aspects – business strategy, organisational structure and competencies – play a major role in achieving the objective stated above. However, these are beyond the scope of this book.

The following section will elaborate the business view and its enablement by technology.

7.2 The business view

We have discussed at length in previous chapters the business processes that an aviation MRO organisation employs to achieve airworthiness of an aircraft. Here we will discuss the view of how the business can intermesh with the technology.

When we view a business from the data, function and workflow perspectives, we open the door for an IT solution to be created. So, what we need is an enterprise data model, which indentifies and describes all the data required to run an aviation MRO business; then we need to identify all the functions that an aviation MRO business performs as well as how they flow into each other.

So what I am saying is that we need an industry data model. Similarly, we need an industry business process model which describes the aviation MRO business using APQC or an equivalent standard.

We will continue with our discussions around IT enablement with an ssumption that there is an aviation MRO industry data model and a business process model. The current state is that there are many proprietary data models, tied to specific solutions, as well as business process models from variety of vendors.

Again there are approximately thirty business scenarios (or use-cases) that an aviation MRO business deploys to achieve the outcomes. These scenarios have been discussed in earlier chapters. The details of these specific scenarios are outside the scope of this book.

7.3 The ideal solution

In order to build something, which is useful and fit for purpose as well as flexible and scalable, one needs a proven framework to start with. Once the framework is identified it is easy to architect a solution, be it a building or a computer system. Until now most of the aviation MRO solutions have been point solution and not necessarily architected. This has been the bane of the industry because it is almost ridiculous to find that the maintenance process of an aircraft, which is so standardised, does not have a standardised IT enabler. Our ideal solution shall make sure that it is built on a standardised platform.

7.3.1 It is all about process

Almost all the software vendors, including SAP and ORACLE, claim that they have a business process model for aviation MRO. And this is true. However, none of them are open and certified by an organisation such as APQC. These proprietary business process models, unfortunately, are more focused towards the internal workings of the software rather than the industry itself. This is evident in the amount of effort spent in implementing MRO software for an organisation. This is because a significant amount of customisation is required before the system becomes acceptable. In addition, every organisation tries to build its own business process model as they launch into either implementation of a new system or business transformation.

The other factor that also comes into play is that all the MRO organisations think that they are unique and they need their own business process model. So they hire consultants or put a team of business analysts to create their own 'unique' model. This always has the risk of creating software, by configuration or by customisation or both, in a straight-jacket. Later on this comes back to bite them as they are unable to scale and the software is not flexible enough to incorporate innovations in technology or changes in the business model due to changes in market conditions.

Today, when a new aircraft with new embedded technologies comes into play, i.e. B787 and A380, it demands a change in core applications.

Therefore, there is a need for a universally accepted business process model, which describes the entire value chain: from receiving an order to delivering a certified aircraft.

This business process model will have processes defined from value chain to activity level. APQC defines the levels in a process model as follows:

1. Level 1: Value chain
2. Level 2: Process areas

3. Level 3: Processes

4. Level 4: Activities.

The way it works is that once all the processes are identified and defined to the lowest possible levels then the business scenarios can be applied to create the required specifications for a solution. The intersection of the business scenarios and processes are the requirements for IT enablement. This is a highly specialised technique and requires a separate discussion altogether.

For the purpose of this book, it suffices to say that a standard business process model and business scenarios can be made available to define what the IT solution should accomplish.

7.3.2 So, the software enables the process ... easy!

The term application software implies a group of programs, which run in an environment using a database. The purpose of this application software is to capture, create, amend, delete and display data in various forms via different user interfaces. We have discussed earlier the types of software that are in use in this industry and their nuances. Therefore we assume, without any bias that there will be a set of applications, either bundled or interoperable, which will be able to deliver information as specified by the business scenarios and business process model; in other words, requirements.

We would certainly like to believe that one single group of software will be sufficient to fulfil the requirements. However, in real life scenarios this is not the case. Whatever we try to do in order to meet the requirements fully we will need multiple applications, which are specifically designed to cater to certain functions. They will need to be loosely coupled so that they can be individually replaced with ease and without any major disruptions when either they become obsolete or the requirements change.

The best part of aviation MRO processes is that they are prescriptive and most are universal because of the rigour of the regulations. Hence, the expected changes are minimal from a process point of view. Today, as long as MSG-3 is applicable the core certification, processes will not change. However, as the technologies change, whether in IT or aircrafts, they will impose changes to the application software.

An ideal set of application software will fit into service oriented architecture, interoperating with each other with ease. This architecture will allow for scaling up and down in functionality to match whatever kind of business the aviation MRO organisation wants to be in. In other words, the set of software should be able support a full blown MRO operation as well as be able to scale down to just line maintenance, if that is what the organisation does.

7.3.3 The infrastructure – where the software will run and data will reside

The technology on which the application software runs is changing at breakneck speed. Notwithstanding the growth in speed of processing and memory capacities, it is the delivery mechanism that is changing as well. Literally, now the IT infrastructure sits in the clouds.

The aviation MRO solution should also sit in the 'cloud'. However, there are issues, which need to be considered and resolved.

The main issue is that none of the current software vendors have prepared themselves for this change and nor do they have the appetite to invest in this. This is besides the issues of privacy, pricing and contract management. The fact that in this industry it is mandatory to keep the history of an aircraft and its component for its entire life cycle – phase-in to phase-out – and also raises the issue of data archiving and transfer when an aircraft changes hand.

There are multiple aspects of the infrastructure: the hardware, network and operating systems. On operating systems there is not much debate nowadays as even mainframes (IBM Z-Series) are also moving to the open standards platform like LINUX. In other words, the most suitable platform for aviation MRO software will be an open standards operating system – UNIX in its various forms.

The network under the internet protocols is ubiquitous and also does not need much discussion. The aviation MRO solution must be delivered on an internet protocol-based network.

As far as the hardware is concerned, nowadays with the advent of 'cloud' computing it does not really matter which hardware is used as long as it provides optimal response, and a high degree of scalability, reliability, openness and flexibility. It is important to note that processing power is one of the critical requirements from the hardware deployed. Because, as we discussed earlier, aviation MRO processes comprise of low volume but very complex of transactions. However, due to regulatory requirements very high volumes of data need to be kept online for a very long period, which should also lend itself to fast queries as and when required.

There are two more aspects of hardware that need to be mentioned here to complete the infrastructure discussion: the ability to interface with many external systems and the ability to use various types of user interface.

An aviation MRO application will never be able to run in isolation, it needs to interact with various internal and external applications and systems. Therefore it should be able to expose itself for easy integration with other systems, and these interfaces should be based on open standards.

Insofar as user interfaces go, the latest technologies have a lot to offer: from hand-held devices to sensors to RFID and the like. The infrastructure should be able to seamlessly accommodate all these various technological marvels.

On the whole, aviation MRO application needs a robust, well architected infrastructure to be able to respond to the market and add value by providing a scalable and flexible platform.

7.3.4 Integrated and responsive solution

With the core application software and infrastructure in place, it is time to discuss how this solution could be made to be responsive with an integrated view of the information.

As we saw in the earlier chapters, in order to enable all the processes a set of application software will be required. A single software package will not be able to achieve the ultimate. There are functions like Finance, Human Resources, Training, Sales, Marketing, etc., which are used to generate different sets of information. In addition to these there are external systems, which provide information, and all these need to be integrated to create a single source of truth. Without such an integrated view of data no meaningful analysis can be conducted.

Therefore there is a need for an integrated view of all the data required or generated by the set of application programs, which enable the aviation MRO business processes. This could be implemented as a data warehouse, which will supply data to any analytical tools thus raising the level of business intelligence of the aviation MRO organisation. With the single source of truth, i.e. the integrated set of data, it will be possible to measure the performance of the organisation and derive the metrics, discussed earlier in this chapter.

With the availability of integrated data and the ability to capture data as and when needed, near-real time if necessary, the aviation MRO organisation will be able to not just quickly analyse the data but also will be able predict and forecast by using advanced analytical tools that are available in the market today. This ability will make them agile and responsive to the changes in the market condition and ensure airworthiness.

The other aspect of responsiveness is about the use of the components of the software. As we discussed earlier, it is this process that drives the software and dictates as to how it will be configured. We have assumed that there is a set of software, which is highly configurable and service oriented. Therefore, if we are able to record the processes in such a tool, which has the ability to influence the configuration of the software, we will have very responsive software indeed. This is because as and when the process changes, either driven by the market conditions or innovations in technology, it can easily reconfigure the set of application software accordingly.

Similarly, one should be able to hook-in any kind of device that is required for data capture and display.

The other major area where technology is spreading its wing is the 'health monitoring' systems. The engine health monitoring systems have been in place for a while. However, it is only now that aircraft health monitoring systems are becoming popular. The purpose of these systems is to constantly monitor the status of the structure, systems and components on the aircraft, whether it is flying or not. This is supposed to eliminate dependence on visual and other non-destructive inspections and tests. These health monitoring systems require their own infrastructure and their details are beyond the scope of this book. However, the infrastructure for aviation MRO organisation, should have the ability to integrate or interface with the health monitoring systems.

7.3.5 Bringing everything together

Let us now see how we can bring all this together. The starting point is a conceptual process model, which can be used to visualise the aviation MRO business.

An aircraft is designed and manufactured by the OEM which documents all the design details and maintenance instructions. These documents are delivered along with the aircraft to the operator. The operator then hands over all the relevant information and documents to the aviation MRO organisation. Now this organisation collects all the necessary data and creates maintenance plans down to Task Cards and their sequencing. There Task Cards are derived from the documentation provided by the OEMs. The task is then executed and reported. Subsequently, the aviation MRO organisation collects and collates all the relevant information and ensures that the aircraft is airworthy by following all the instructions and guidelines provided to it. In other words an aviation MRO organisation generates its revenue by getting the orders to maintain the aircraft, conducting engineering design and planning with the intention of managing the configuration of the aircraft, provisioning tools and material and executing the maintenance tasks so that the aircraft is certified to fly again: the aircraft is made airworthy.

So the enabling IT captures and manages the technical documentation, which is used for creating Task Cards and providing support for maintenance execution; helps plan maintenance activities and keeps track of the removal and installation of components from and to the aircraft; and stores the data in a data warehouse for analytical purposes where the processes are maintained as current as possible to reconfigure the software as and when required.

All this can be done by applying the right technologies and integration techniques.

7.4 Is this feasible?

Of course it is!

In this chapter we discussed at a very high level what is required to create an all-singing-and-dancing IT system to enable an aviation MRO organisation attain airworthiness. Not just that the aircraft is safe enough to fly, but that it is also worth flying for the operators.

So what do we need? We need the following, which can become aviation MRO industry standards:

1. Business Process Model

2. Enterprise Data Model

3. Reference Architecture

   - Business
   - Application
   - Technology.

Once these standards are accepted and published, any software that complies with them will do the job. This is not rocket science. All this has been institutionalised by TOGAF (The Open Group Architecture Framework).

Then there is a question of investment. Who will invest in building these? The software companies have still to decide. However, some of the companies already possess some these assets but none of them have all the assets required. Maybe one day one of them will take the lead and dash forth to capture the market.

IBM has an asset, the MRO Business Process model, which goes down to level 3 (process). However, it is not APQC certified and has not been accepted by the industry as universal. Similarly, SAP has one. However, building a comprehensive business process will require the backing of regulatory rganisations and the coming together of all the stakeholders. ATA does have an initiative and it will probably lead the way as far as business process modelling is concerned.

Unfortunately there is no MRO specific industry enterprise data model which is universally accepted. However, there are plenty of bits of logical model with many vendors and even MRO companies. These can be brought together in a cooperative manner and established with a comprehensive conceptual and logical data model.

There are various architectures that are being touted for aviation MRO; but none have yet been accepted by the industry as reference architecture. This again requires a collaborative effort and can be achieved in a very short time. IBM proposes one of the architectures, shown in Figure 7.2, which could become the starting point.

One of the main points to keep in mind is that although the aviation MRO business scenarios are quite standardised in practice, they have not been fully documented and approved as industry standards as a whole. The FAA and others have specified the requirements for the outcome of these processes but not the processes themselves, even though they are implied. Therefore, a comprehensive set of business scenarios need to be developed and certified as industry standards or even best practices. Since almost all vendors profess to have a set of best practices, this task should not be difficult.

Figure 7.2 IBM architecture

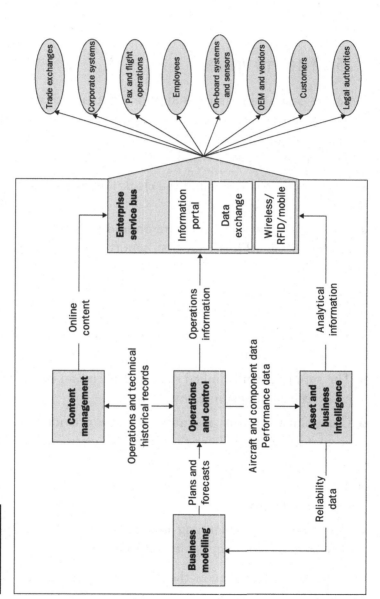

Once all the above is in place, as stated before, a set of software running on an appropriate platform is very feasible.

7.5 The nirvana

Here we start with the assumption that an ideal set of software and IT infrastructure exists, which meets all the requirements of an aviation MRO organisation. Then this is what happens in the ideal wonderland of aircraft maintenance:

- The aircraft records round the clock how it feels through health monitoring systems.
- The parts show when they are on and when they off the aircraft using RFID. They also report their whereabouts and condition.
- The technicians have supporting documentation for fault rectification and reporting using mobile devices.
- The demand for replacement parts can be predicted using complex algorithms.
- The supply chain is linked such that parts can be ordered and received almost instantly.
- Compliance reports are automatically provided to the regulators and they have the ability to query information themselves.
- Technical documentation is received, managed and distributed automatically.
- The maintenance planning is automated and is able to take last minute changes into account.

Some of these activities in the ideal world are depicted in Figure 7.3.

Figure 7.3 Some activities of an ideal infrastructure

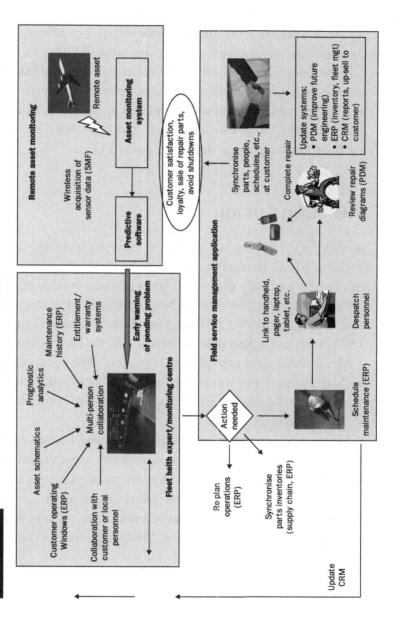

Remote asset monitoring

Remote asset

Wireless acquisition of sensor data (SMF)

Asset monitoring system

Predictive software

Asset schematics

Prognostic analytics

Maintenance history (ERP)

Customer operating Windows (ERP)

Entitlement/ warranty systems

Multi-person collaboration

Collaboration with customer or local personnel

Fleet helth expert/monitoring centre

Early warning of pending problem

Customer satisfaction, loyalty, sale of repair parts, avoid shutdowns

Field service management application

Action needed

Re-plan operations (ERP)

Synchronise parts inventories (supply chain, ERP)

Schedule maintenance (ERP)

Link to handheld, pager, laptop, tablet, etc.

Despatch personnel

Synchronise parts, people, schedules, etc., at customer

Complete repair

Review repair diagrams (PDM)

Update systems:
• PDM (improve future engineering)
• ERP (inventory, fleet mgt)
• CRM (reports, up-sell to customer)

Update CRM

As stated earlier, an aviation MRO system has very few transactions as the aim of the ultimate system is to try and altogether eliminate data entry by people. The current day technology of sensors and RFID can help achieve that. The aviation MRO managers and staff should use the systems for decision making, reference for expertise, and receiving instructions. This will be nirvana for the aviation MRO organisation, where the right skills are used for the right purpose with the support of IT.

7.6 Conclusion

In this final chapter we discussed how to leverage IT for aviation MRO by specifying an ideal system. However, the technology can only go so far. An organisation does not just run on technology, but the technology enables the managers and the workforce to run the organisation most efficiently. The system envisaged in this chapter will do exactly that.

Meanwhile, the purpose of this chapter was not to design an MRO system but to provoke thoughts as to what kind of system can be designed and developed. We have remained at a conceptual level.

This is not say that we should not try and find out what benefits an aviation MRO organisation can expect from such an ideal system. Figure 7.4 is an attempt to set an expectation for quantified benefits in specific areas. These are achievable benefits, and hopefully one day we will have such an MRO system that will help an aviation MRO organisation achieve these.

| Figure 7.4 | Expectations for an ideal MRO system |

- Materials Management
    - Reduce indirect headcount 10%
    - Reduce component shop direct headcount by 5 to 10%
    - 10%–15% reduction in rotables and repairables inventory

- Maintenance Record Management
    - Reduction of 1 FTE per 25 aircraft required for collation and arching of paper documents
    - Elimination of 8–15 staff months of effort needed to retrieve and collate records at aircraft phase-out

- Operational and Technical Information Management
    - 10%–15% increase in direct hangar productivity,
    - 20%–30% reduction in indirect labour required for management of document evaluation and distribution

- Maintenance Management Business
    - Reduce indirect headcount by 10 to 15%
    - Reduce aircraft related direct headcount by 15 to 20%
    - 1%–2% increase in hull availability based on utilising improved productivity to reduce downtime
    - 20% improvement in hangar and workshop utilisation

7.7 Notes

1. Isa Upanishad, *Upanishads*, Oxford University Press; translated from Sanskrit by Patrick Olivelle; 1996.
2. *Oxford English Dictionary*, Oxford University Press; 2010.
3. *http://www.answers.com/topic/technical-dispatch-reliability*
4. *http://www.answers.com/topic/aircraft-utilization-rate*

8

Conclusion

Abstract: This chapter summarises the journey that author took to reach here. This chapter highlights the lessons learnt and the trials and tribulations of writing a book of this kind.

Keywords: experience, location, time management, mega centre, additive manufacturing.

If . . . the past may be no rule for the future, all experience becomes useless and can give rise to no inference or conclusion.[1]

As with many people, I have often been told by friends and colleagues to 'write a book'. Whether this was a polite way to ask me to get off my soapbox, or a genuine admiration for my accumulated knowledge, theories and ideas, we may never know.

I know that during my career in aviation MRO, I often wished that there was more published information on how software could be leveraged to revolutionise what is a highly regulated process, in order to bring to this practice the efficiencies that an ERP system, say, brought to the financial processes.

Over the years of attending seminars, conducting and participating in workshops, and talking to customers and clients who all wanted to understand if technology could be leveraged to support the processes and had all struggled in their own way to make sense of what was on offer, it slowly dawned on me that I did in fact have enough information inside my head to actually write a book dealing with this complex and unique subject.

I would like to say that all the accumulated knowledge of many years in the airline business, and as a consultant to airlines, flowed smoothly and without impediment to the virtual page in my computer, but in reality it took me nearly two years to complete this slim book. While far from an easy subject, the writing of this book was a way for me to imagine the future and to build in my mind a possible view of the MRO organisation of the future.

In recent years, at many a seminar and workshop around the globe, the idea of a mega-aviation-MRO centre has been discussed in stronger and louder terms than ever before. To some, this is still farfetched and almost impossible to achieve; however, I believe this is a serious discussion and deserving of greater scrutiny.

The question of a mega-centre capable of handling thousands of aircraft, and operating a little bit like WalMart in terms of automation and cost control, may be difficult to imagine, however the concept is valid. Aircraft maintenance is such a standardised process that centralisation is the most obvious method of driving its efficiencies to its extreme limits. So, even though a physically centralised mega-centre may not be viable a virtualised mega-centre is possible.

Imagine, if you will, a massive MRO facility, which is strategically located somewhere on the globe and is accessible to all the airlines in the world. This facility services all the aircraft for their heavy maintenance needs, stocks all the necessary spares parts, and is able to schedule and execute all the heavy checks for all types of aircraft. The aircraft from various airlines fly in dead-heading to the mega-centre whenever their maintenance is scheduled. A magnificent scenario, indeed! This is because the mega-centre will be able to optimise all the aspects of aircraft maintenance, as described earlier. However, there is a snag. This will require acres and acres of land and a massive infrastructure to be able to service thousands of aircraft at a time. This is will also require a gargantuan logistics planning and execution; to say nothing of the amount of investment that will be needed to acquire land and build and maintain these facilities.

However, let us say a set of processes are developed, which are based on such a hypothetical mega-centre and a set of IT enablers are developed to support these processes, then there is a fair possibility of creating a virtual mega-centre, I believe. A federated business model, which can be franchised, could be developed and put in place that comes with the all the business functions pre-defined and packaged and can be implemented in various locations. Thus solving the problems related to size, logistics, supply-chain, as well as investment. All the franchised units will function like a clone with only a limited customisation.

Franchising has been defined in many ways but the one that is most apt in this context is that of a method to quickly and cost effectively deploy a mature and standardised business process at various locations, without significantly localising it. aviation MRO processes are standardised and mature. However, there hardly has been an attempt to exploit the 'franchising' concept. There are many reasons why this has not happened. Many of those are related to the issues discussed previously in this book. They collectively point to one, and only one, reason though: there is no MacDonald of the aviation MRO. Maybe this is what the industry is waiting for and needs. Whether it will be Boeing, Rolls Royce or

Bombardier, who have all at least toyed with this idea, or someone else, is to be seen.

The key to an aviation MRO franchise will be information technology. Not just the technology available today but also the ones which make additive manufacturing possible. The 3-D Printing is a disruptive technology but one that has the potential to bring about a revolutionary change in the way an aviation MRO organisation procures a part. This could eliminate the need to stock the MSI parts but just print one on demand. This technology is not there yet, but there is a huge possibility that it could become the norm in the future. But this is not to say that it will not happen.

Similarly RFID combined with mobility can introduce an unbelievable amount of efficiency, not seen before. These technologies are still evolving and their applications have yet to cross practical and regulatory hurdles. There are many unknowns accompanied by myths, especially the ones about 'interference' with navigational systems of an aircraft. There may be some truth in the fact that electronic wave-emitting instruments like mobile phones could interfere with the navigational equipment of the aircraft. That is why, I suppose, millions of man-hours are spent by passengers switching the mobile phones on and off, and cabin crew making sure that they are switched off. This is a correct and valid procedure but only when the design of the navigational instruments are constant. What if these instruments are designed to be interference proof? Then information technology solutions like RFID and mobility can be widely deployed. There has been significant development in this area and shortly and surely the aviation MRO organisations will leverage these capabilities.

In the reality of today, every MRO service provider is struggling with its information systems. They are either trying to enhance their systems or replace them. However, it is my belief that what is required is to envision a new type of organisation that is facilitated and supported by information technology.

In conclusion I would say that the balancing of safety and cost effectiveness in the aviation MRO industry can only come about with using disruptive innovations and leveraging information technology. I shall quote Charles Darwin, in support of my conclusion: 'I am convinced that natural selection has been the main but not the exclusive means of modification.'[2]

The Eldorado of aviation MRO could be a set of federated facilities, utilising standardised processes made efficient by information technology.

Notes

1. David Hume: *An Enquiry Concerning Human Understanding* (1748), ed. L. A. Selby-Bigge (1894), section 4, part 2, 37–8.
2. In *The Origin of Species by Means of Natural Selection* with additions and corrections from sixth and last English edition (1899), Vol. 2, 293.

Appendix

Airworthiness definitions

This book uses the term 'airworthiness' repeatedly because the aim of an aviation MRO organisation is to achieve airworthiness for an aircraft. However, this term has been defined by many authoritative organisations and the following are the examples of such definitions.

According to the FAA the definition of airworthiness is: 'an aircraft is airworthy when it has been given a certificate of airworthiness by a competent authority'. The conditions (abbreviated) are described below:

> An applicant for a standard airworthiness certificate for aircraft ... is entitled to a standard airworthiness certificate if –
>
> 1 He presents evidence to the administrator that the aircraft conforms to a type design approved under a type certificate or a supplemental type certificate and to applicable Airworthiness Directives;
> 2 The aircraft (except an experimentally certificated aircraft that previously had been issued a different airworthiness certificate under this section) has been inspected in accordance with the performance rules for 100-hour inspections set forth in part 43.15 of this chapter and found airworthy by—

      a. The manufacturer;

      b. The holder of a repair station certificate as provided in part 145 of this chapter;

      c. The holder of a mechanic certificate as authorized in part 65 of this chapter;

      d. The holder of a certificate issued under part 121 of this chapter, and having a maintenance and inspection organization appropriate to the aircraft type; and

3 The Administrator finds after inspection, that the aircraft conforms to the type design, and is in condition for safe operation.

A more generic and non-process orientated definition for airworthiness is in JSP553 Military Airworthiness Regulations (2006) Edition 1 Change 5. It says:

> The ability of an aircraft or other airborne equipment or system to operate without significant hazard to aircrew, ground crew, passengers (where relevant) or to the general public over which such airborne systems are flown.

This definition applies equally to civil and military aircraft.

An example of a method used to delineate 'significant hazard' is a risk reduction technique used by the military and used widely throughout engineering known as ALARP (As Low as Reasonably Practicable). This is defined as:

The principal, used in the application of the Health and Safety at Work Act, that safety should be improved beyond the baseline criteria so far as is reasonably practicable. A risk is ALARP when it has been demonstrated that the cost of any further Risk reduction, where cost includes the loss of capability as well as financial or other resource costs, is grossly disproportionate to the benefit obtained from that Risk reduction.

In Canada, *Canadian Aviation Regulations*, CAR 101.01, Subpart 1: – Interpretation Content last revised: 2007/12/30

> 'Airworthy' – in respect of an aeronautical product, means in a fit and safe state for flight and in conformity with its type design.

Airworthiness certification

The order 8130.3G by the FAA on 31 August 2010, deals with airworthiness certification of aircraft and related products. This is a national policy and is publicly available.

> This order establishes procedures for accomplishing original and recurrent airworthiness certification of aircraft and related products and articles. The procedures contained in this order apply to Federal Aviation Administration (FAA) manufacturing aviation safety inspectors (ASI), to FAA airworthiness ASIs, and to private persons or organizations delegated authority to issue airworthiness certificates and related approvals.

This order is 318 pages long, describing almost all the aspects of airworthiness certification of a new aircraft, special aircraft and to aircraft going out of service. This comprehensive order has eight chapters and seven appendices. It is a mandatory reading for practitioners of aircraft maintenance. In order to get a brief idea of what it contains, the chapters are listed below:

Chapter 1. Introduction

Chapter 2. General Policies and Procedures
Section 1. General Information
Section 2. Airworthiness Certificates
Section 3. Initial or Subsequent Issuance of Airworthiness
Certificates (Original/Recurrent) or Related Approvals

Chapter 3. Standard Airworthiness Certification
Section 1. General Information
Section 2. New Aircraft
Section 3. Used Aircraft and Surplus Aircraft of the U.S.
Armed Forces

Chapter 4. Special Airworthiness Certification
Section 1. General Information
Section 2. Restricted Airworthiness Certification
Section 3. Multiple Airworthiness Certificates
Section 4. Limited Airworthiness Certification
Section 5. Primary Category Aircraft Airworthiness
Certifications
Section 6. Light-Sport Aircraft (LSA) Category Aircraft
Airworthiness Certifications
Section 7. General Experimental Airworthiness
Certifications
Section 8. Experimental LSA Airworthiness Certifications
Section 9. Experimental Amateur-Built Airworthiness
Certifications
Section 10. Certification and Operation of Aircraft under
the Experimental Purpose(s) of Exhibition and Air
Racing
Section 11. Certification and Operation of Aircraft under
the Experimental Purpose(s) of Research and
Development, Showing Compliance with Regulations,
Crew Training, Market Surveys, and Operating Kit-
Built Aircraft

Section 12. Provisional Airworthiness Certification
Section 13. Special Flight Permits

Chapter 5. Export Approval Procedures
Section 1. General Information
Section 2. Export Approvals

Chapter 6. Import Procedures
Section 1. General Information
Section 2. Import Aircraft
Section 3. Aircraft Engines, Propellers, and Articles

Chapter 7. Special Flight Authorizations (SFA) for Non-U.S.-Registered Civil Aircraft

Chapter 8. Processing Forms, Reports, and Certification Files

A sample

FAA FORM 8130-6, APPLICATION FOR U.S. AIRWORTHINESS CERTIFICATE (see pp. 226–8)

U.S. Department
of Transportation
**Federal Aviation
Administration**

APPLICATION FOR U.S. AIRWORTHINESS CERTIFICATE

─────────────── **Privacy Act Statement** ───────────────

Information on FAA Form 8130-6, Application for U.S. Airworthiness Certificate is solicited under the authority of 49 U.S.C. 44103 as implemented by 14 CFR Part 21. The purpose of this information is to evaluate an applicant's application for a U.S. Airworthiness Certificate. Submission of this data is mandatory and will become part of the Privacy Act system of records DOT/FAA 801, Aircraft Registration System. Incomplete submission may result in delay or denial of your request. Information maintained in the Privacy Act system of records is routinely used to (1) provide aircraft owners and operators information about potential mechanical defects or unsafe conditions of their aircraft in the form of airworthiness directives, (2) locate specific individuals or aircraft for accident investigation, violation, or safety related requirements, (3) prepare an Aircraft Registry in magnetic tape and microfiche form as required by ICAO agreement, containing information on aircraft owners by name, address, United States Registration Number, and type of aircraft, and (4) DOT Prefatory Statement of General Routine Uses.

Paperwork Reduction Act Statement:
This information is collected for the purpose of issuing a U.S. Airworthiness Certificate to any applicant meeting the criteria established in FAA Order 8130.2, Airworthiness Certification of Aircraft and Related Products. The FAA uses the information to maintain and update the current database of aircraft having obtained approved airworthiness certificates. The burden associated with completing FAA Form 8130-6 is 42 minutes. Providing this information is mandatory if an applicant wishes to obtain an airworthiness certificate. The information is protected under the provisions of the Privacy Act and the Privacy Act system of records DOT/FAA-801, Aircraft Registration System. An agency may not conduct or sponsor and a person is not required to respond to a collection of information unless it displays a currently valid OMB control number. The OMB control number associated with this collection of information is 2120-0018.

TEAR OFF THIS COVER SHEET BEFORE SUBMITTING THIS FORM

FAA Form 8130-6 (4/11) All Previous Editions Superseded

FAA FORM 8130-6, APPLICATION FOR U.S. AIRWORTHINESS CERTIFICATE

Form Approved O.M.B. No. 2120-0018
Expiration Date 02/28/2013

U.S. Department of Transportation
Federal Aviation Administration

APPLICATION FOR U.S. AIRWORTHINESS CERTIFICATE

INSTRUCTIONS - Print or type. Do not write in shaded areas; these are for FAA use only. Submit original only to an authorized FAA Representative. If additional space is required, use attachment. For special flight permits complete Sections II, VI, and VII as applicable.

I. AIRCRAFT DESCRIPTION

1. REGISTRATION MARK	2. AIRCRAFT BUILDER'S NAME (Make)	3. AIRCRAFT MODEL DESIGNATION	4. YR. MFR.	FAA CODING
5. AIRCRAFT SERIAL NO.	6. ENGINE BUILDER'S NAME (Make)	7. ENGINE MODEL DESIGNATION		
8. NUMBER OF ENGINES	9. PROPELLER BUILDER'S NAME (Make)	10. PROPELLER MODEL DESIGNATION		11. AIRCRAFT IS (Check if applicable) IMPORT

II. CERTIFICATION REQUESTED

APPLICATION IS HEREBY MADE FOR: (Check applicable items)

A	1	STANDARD AIRWORTHINESS CERTIFICATE (Indicate category)		NORMAL	UTILITY	ACROBATIC	TRANSPORT	COMMUTER	BALLOON	OTHER

B		SPECIAL AIRWORTHINESS CERTIFICATE (Check appropriate items)								
	7	PRIMARY								
	9	LIGHT-SPORT (Indicate Class)		Airplane	Power-Parachute	Weight-Shift-Control	Glider	Lighter than Air		
	2	LIMITED								

	5	PROVISIONAL (Indicate class)	1	CLASS I						
			2	CLASS II						

	3	RESTRICTED (Indicate operation(s) to be conducted)	1	AGRICULTURE AND PEST CONTROL	2	AERIAL SURVEY	3	AERIAL ADVERTISING
			4	FOREST (Wildlife conservation)	5	PATROLLING	6	WEATHER CONTROL
			0	OTHER (Specify)				

	4	EXPERIMENTAL (Indicate operation(s) to be conducted)	1	RESEARCH AND DEVELOPMENT			2	AMATEUR BUILT	3	EXHIBITION
			4	AIR RACING			5	CREW TRAINING	6	MARKET SURVEY
			0	TO SHOW COMPLIANCE WITH THE CFR			7	OPERATING (Primary Category) KIT BUILT AIRCRAFT		
			8	OPERATING LIGHT-SPORT	8A	Existing aircraft without an airworthiness certificate & do not meet § 103.1				
					8B	Operating Light-Sport Kit-built				
					8C	Operating light-sport previously issued special light-sport category airworthiness certificate under § 21.190				
			9	UNMANNED AIRCRAFT	9A	RESEARCH AND DEVELOPMENT		9C	CREW TRAINING	
					9B	MARKET SURVEY				

	8	SPECIAL FLIGHT PERMIT (Indicate operation to be conducted, then complete Section VI or VII as applicable on reverse side)	1	FERRY FLIGHT FOR REPAIRS, ALTERATIONS, MAINTENANCE, OR STORAGE		
			2	EVACUATE FROM AREA OF IMPENDING DANGER		
			3	OPERATION IN EXCESS OF MAXIMUM CERTIFICATED TAKE-OFF WEIGHT		
			4	DELIVERING OR EXPORTING	5	PRODUCTION FLIGHT TESTING
			6	CUSTOMER DEMONSTRATION FLIGHTS		

C	6	MULTIPLE AIRWORTHINESS CERTIFICATE (Check ABOVE "Restricted Operation" and "Standard" or "Limited" as applicable)

III. OWNER'S CERTIFICATION

A. REGISTERED OWNER (As shown on certificate of aircraft registration) — IF DEALER, CHECK HERE ➤

NAME	ADDRESS

B. AIRCRAFT CERTIFICATION BASIS (Check applicable blocks and complete items as indicated)

AIRCRAFT SPECIFICATION OR TYPE CERTIFICATE DATA SHEET (Give No. and Revision No.)	AIRWORTHINESS DIRECTIVES (Check if all applicable ADs are complied with and give the number of the last AD SUPPLEMENT available in the biweekly series as of the date of application)
AIRCRAFT LISTING (Give page number(s))	SUPPLEMENTAL TYPE CERTIFICATE (List number of each STC incorporated)

C. AIRCRAFT OPERATION AND MAINTENANCE RECORDS

CHECK IF RECORDS IN COMPLIANCE WITH 14 CFR section 91.417	TOTAL AIRFRAME HOURS		3	EXPERIMENTAL ONLY (Enter hours flown since last certificate issued or renewed)

D. CERTIFICATION - I hereby certify that I am the registered owner (or his agent) of the aircraft described above, that the aircraft is registered with the Federal Aviation Administration in accordance with Title 49 of the United States Code 44101 et seq. and applicable Federal Aviation Regulations, and that the aircraft has been inspected and is airworthy and eligible for the airworthiness certificate requested.

DATE OF APPLICATION	NAME AND TITLE (Print or type)	SIGNATURE

IV. INSPECTION AGENCY VERIFICATION

A. THE AIRCRAFT DESCRIBED ABOVE HAS BEEN INSPECTED AND FOUND AIRWORTHY BY: (Complete the section only if 14 CFR part 21.183(d) applies)

2	14 CFR part 121 CERTIFICATE HOLDER (Give Certificate No.)	3	CERTIFICATED MECHANIC (Give Certificate No.)	6	CERTIFICATED REPAIR STATION (Give Certificate No.)
5	AIRCRAFT MANUFACTURER (Give name or firm)				

DATE	TITLE	SIGNATURE

V. FAA REPRESENTATIVE CERTIFICATION

(Check ALL applicable block items A and B)

A. I find that the aircraft described in Section I or VII meets requirements for		THE CERTIFICATE REQUESTED
	4	AMENDMENT OR MODIFICATION OF CURRENT AIRWORTHINESS CERTIFICATE

B. Inspection for a special flight permit under Section VII was conducted by:	FAA INSPECTOR		FAA DESIGNEE		
	CERTIFICATE HOLDER UNDER		14 CFR part 65	14 CFR part 121 OR 135	14 CFR part 145

DATE	MIDO/FSDO OFFICE		FAA INSPECTOR'S SIGNATURE or DESIGNEE'S SIGNATURE AND NO.	FAA INSPECTOR'S CERTIFICATION FILE REVIEW SIGNATURE
		4		1

FAA Form 8130-6 (4/11) All Previous Editions Superseded   Electronic Format --PDF   Page 1 of 2

	A. MANUFACTURER		
VI. PRODUCTION FLIGHT TESTING	NAME		ADDRESS
	B. PRODUCTION BASIS *(Check applicable item)*		
		PRODUCTION CERTIFICATE *(Give production certificate number)* ▶	
		TYPE CERTIFICATE	
		OTHER:	
	C. GIVE QUANTITY OF CERTIFICATES REQUIRED FOR OPERATING NEEDS		
	DATE OF APPLICATION	NAME AND TITLE *(Print or type)*	SIGNATURE

	A. DESCRIPTION OF AIRCRAFT		
VII. SPECIAL FLIGHT PERMIT PURPOSES OTHER THAN PRODUCTION FLIGHT TEST	REGISTERED OWNER		ADDRESS
	BUILDER *(Make)*		MODEL
	SERIAL NUMBER		REGISTRATION MARK
	B. DESCRIPTION OF FLIGHT	CUSTOMER DEMONSTRATION FLIGHTS ☐ *(Check if applicable)*	
	FROM		TO
	VIA		DEPARTURE DATE	DURATION
	C. CREW REQUIRED TO OPERATE THE AIRCRAFT AND ITS EQUIPMENT		
	PILOT	CO-PILOT	FLIGHT ENGINEER	OTHER *(Specify)*

D. THE AIRCRAFT DOES NOT MEET THE APPLICABLE AIRWORTHINESS REQUIREMENTS AS FOLLOWS:

E. THE FOLLOWING RESTRICTIONS ARE CONSIDERED NECESSARY FOR SAFE OPERATION: *(Use attachment if necessary)*

F. CERTIFICATION - I hereby certify that I am the registered owner (or his agent) of the aircraft described above; that the aircraft is registered with the Federal Aviation Administration in accordance with Title 49 of the United States Code 44101 *et seq.* and applicable Federal Aviation Regulations; and that the aircraft has been inspected and is safe for the flight described.

DATE	NAME AND TITLE *(Print or type)*	SIGNATURE

VIII. AIRWORTHINESS DOCUMENTATION (FAA/DESIGNEE use only)	A. Operating Limitations and Markings in Compliance With 14 CFR Section 91.9, As Applicable	G. Statement of Conformity, FAA Form 8130-9 *(Attach when required)*
	B. Current Operating Limitations Attached	H. Foreign Airworthiness Certification for Import Aircraft *(Attach when required)*
	C. Data, Drawings, Photographs, etc. *(Attach when required)*	I. Previous Airworthiness Certificate Issued in Accordance With 14 CFR Section _____ CAR _____ *(Original attached)*
	D. Current Weight and Balance Information Available in Aircraft	
	E. Major Repair and Alteration, FAA Form 337 *(Attach when required)*	J. Current Airworthiness Certificate Issued in Accordance With 14 CFR Section _____ *(Copy attached)*
	F. This inspection Recorded in Aircraft Records	K. Light-Sport Aircraft Statement of Compliance, FAA Form 8130-15 *(Attach when required)*

Bibliography

ATA MSG-3: *Operator/Manufacturer Scheduled Maintenance Development*; Revision 2003.1; Copyright © 2003 Air Transport Association of America, Inc.; Air Transport Association of America, Inc. 1301 Pennsylvania Avenue, NW – Suite 1100, Washington, DC> 20004–1707, USA.

AC 120–16D DATE 03/20/03 Initiated by: AFS-300; *Air Carrier Maintenance Programs*; U.S. Department of transportation, Federal Aviation Administration, Flight Standards Service, Washington, D.C.

AC 121–22A DATE: 3/7/97; *Maintenance Review Board Procedures*; U.S. Department of Transportation, Federal Aviation Administration, Flight Standards Service, Washington, D.C.

DOT/FAA/AR-03/70: *Continuing Analysis and Surveillance System (CASS) Description and Models*; Office of Aviation Research, Washington, D.C. 20591. This document is available to the U.S. public through the National Technical Information Service (NTIS), Springfield, Virginia 22161.

ORDER 8130.2F CHG 5, 1/15/2010, U.S. Department of Transportation Federal Aviation Administration; National Policy: Airworthiness Certification of Aircraft and Related Products; Initiated by: AIR-200.

Aviation Week Executive Roundtable: *MRO IT: Extracting MRO Intelligence from a Data-Intensive Aircraft*;

written by Helen Kang, October 15, 2010; *www. aviationweek.com*

MRO IT Market, Suppliers Survey; Issue No. 56, February/March 2008; *Aircraft Commerce*; *www. aircraft-commerce.com*

Travel Terminology Glossary: Compiled by Ken Burch: *Ken_ burch@us.ibm.com*

CAP 642: Airside Safety Management; © Civil Aviation Authority 2005; ISBN 0 11790 646 8.

Data Modeling Essentials: Graeme C. Simsion, Graham C. Witt; Morgan Kaufmann, Nov 4, 2004.

TOGAF 9.1: Copyright © 1999–2011 The Open Group; *http://pubs.opengroup.org/architecture/togaf9-doc/ arch/*

Recommended Reading

Air Carrier MRO Handbook, by Jack Hessburg; Copyright © 2001 by McGraw-Hill Companies. ISBN 0-07-136133-2

Index

ABC analysis, 84
Abu Dhabi Aerospace
    Technologies (ADAT), 179
AC 120-16D, see standards
AC 120-17A, see standards
acceptance, 41–2
accounts receivable (AR), 114
ADABAS, see database software
AeroMexico, 158
Air Canada, 183
Air India, 172, 184
Air Mobility Command (AMC),
    92
Air Transport Association (ATA),
    3, 37–8, 43, 83, 115, 116–7,
    132, 153
Airbus, 94, 119
Aircraft Communications
    Addressing and Reporting
    System (ACARS), 134
aircraft configuration, 57–8
aircraft maintenance, 1–14, 20–23,
    33–113
  business of, 15–31
  hangar/heavy, 21–2
  integrated fleet, 22
  line/ramp, 22
  MRO, see Maintenance, Repair
    and Overhaul
  objectives, 7–9
  organisations, 11–12

  overview, 1–14
  paradigm, 33–113
  planning document/database
    software (MPD), 5, 19, 43,
    45, 69
  planning review board (MPRB),
    96
  planning working group
    (MPWG), 133
  process, 3–7
  resource planning (MRP), 106,
    130, 131, 140, 152, 166, 167
  review board (MRB), 3, 5, 12,
    13
  strategies, 9–11
Aircraft Maintenance and Task
    Oriented Support System
    (AMTOSS), 33, 43, 66–7, 73,
    76–7, 92, 132, 133
aircraft maintenance manual
    (AMM), 43, 45
aircraft on ground (AOG), 2, 16,
    87, 141
airframe, 4, 17, 23, 26, 27, 36, 39,
    42, 46, 51, 53–5, 62, 91
airworthiness, 2–3, 5, 9–10, 13, 16,
    20, 29, 30, 33, 48–51, 62–4,
    95, 98, 105, 108, 115, 123
  certificate, 41, 62, 64, 73
  maintaining, 50–1
  standards, 116

Airworthiness Directive (AD), 56, 98
algorithms, 79, 125
Alitalia, 142, 153, 155
American Productivity and Quality Center (APQC), 34, 35, 113, 193, 199, 200, 208
AMOS, see application software
Ansett, 156
Application Programming Interface (API), 148
application software, 29, 30, 55, 65, 72, 76, 101, 170, 182, 201–2, 204–5
    AMOS, 171, 183, 187
    Complex MRO (CMRO), 169
    EMPACS, 153, 158–9, 179
    EMSYS, 163
    ICS/DL/1, 145
    IMRO, 169
    MAXIMO, 167, 170
    MAXIS, 153
    MAXI-MERLIN, 151, 156-8
    MEMIS, 151, 153, 155, 160, 179
    MERLIN, 139, 151, 153, 155–6, 160, 163, 173, 179–80
    Raptor, 170
    SABRE, 139, 140–1, 158–9, 164, 173
    SCEPTRE, 151, 153–5, 160, 179–80
    TRAX, 158, 171, 173, 187
approval, 41, 60, 96, 105–6, 114
architecture, 34, 35, 36, 42, 103, 115, 149, 150, 151, 166, 177, 178

Systems Network (SNA), 142, 148, 152
systems oriented, 190
ARPANET, 146
ASSEMBLER, see programming language
assembly, 47, 66, 70, 94, 102
asset, 52, 81, 84, 86–4, 101–2
axon, 169

batch, 144, 149
Boeing, 7, 35, 36–7, 43, 94, 117–8, 139, 167–8, 172, 175
Bombardier, 218
bonded warehouse, 81–2, 113
British Airways, 35, 142, 153, 158, 168, 185
buyer furnished equipment (BFE), 120

carriers, 12, 18, 97–8, 104, 108, 114
Caterpillar Tractor, 145
Cathay Pacific, 153, 158, 159
Certification, 40–1
CICS, see communications software
Civil Aeronautics Administration (CAA), 3, 38, 105
COBOL, see programming language
CODASYL, see database software
Code of Federal Regulations (CFR), 97
communications software, 142, 147–8, 154,
    CICS, 142, 146, 147
    IMS DC, 142
    VTAM, 147, 148, 152

Complex MRO (CMRO), see application software
compliance, 4, 7, 12, 35, 52, 54, 65–6, 72, 76, 92, 95, 98, 101
component maintenance manual (CMM), 43, 73, 74, 75
computer, 6–7, 100, 119, 140–7, 148–50, 153, 181
configuration drawing list (CDL), 47
Continuing Analysis and Surveillance System (CASS), 97–9, 114
continuous improvements, 95–7
Crossair, 171
customer care, 30–1
Customer Relationship Management (CRM), 15, 31, 104

database software, 5, 41, 104, 145–6, 147, 152, 154, 156–7, 158, 161–2, 185, 189
  ADABAS, 145, 146, 151, 154, 155, 159
  CODASYL, 143, 145
DBMS, 145–6
IMS DB, 141, 142, 146–7, 151, 154–5, 156, 159, 163, 180
VSAM, 180
DBMS, see database software
Department of Defence (DoD), 43, 48, 118
depreciation, 88
Directorate of Maintenance (DOM), 108, 110
documentation, 39, 42–4, 45, 67, 85, 118, 134, 144
  ATA 2100 standard, 43

iSPEC2200 standard, 43
S1000D standard, 43
DOS, see operating system

Electronic Data Interchange (EDI), 116
EgyptAir, 158
El Al, 184
Emirates, 159, 164
EMPACS, see application software
EMSYS see application software
emulation, 148
EN 54, see standards
engine manual (EM), 43, 45
engine, 66–8
Enterprise Resource Planning (ERP), 118, 130–1, 158, 166–7, 169, 172, 177, 178, 181
  integrated solutions, 184–6
Environmental Protection Agency (EPA), 95
European Aviation Safety Agency (EASA), 92

facilities, 93–5
fatigue, 125
fault identification manual (FIM), 44, 45, 66, 73, 74
fault reporting manual (FRM), 44–5
Federal Aviation Authority (FAA), 3, 11, 15, 17, 41, 91, 96–8, 105–10, 114, 130, 179
Federal Aviation Regulations (FAR), 13, 92, 106
finance, 86–90, 110, 111, 112, 133
forecasting, 124–8

FORTRAN, see programming
    language
freight, 85

Goldfarb, 118
governance, 13, 35, 115
ground support equipment/fleet
    (GSE/F), 36, 39, 51, 54–5, 64,
    75, 76–8, 90, 94, 110, 112,
    125
ground support facilities (GSF),
    110, 112
Gulf Air, 154
Gulf Aircraft Maintenance
    Company (GAMCO), 179–80

hardware, 97, 100–2, 144, 146,
    154, 171
Hewlett Packard (HP), 170
Hindustan Computer Limited
    (HCL), 169
Hong Kong Aircraft Engineering
    Co Ltd (HAECO), 179
HTML, see programming
    language
human resources, 90–3, 101, 133

ICS/DL/1, see application software
illustrated parts catalogue (IPC),
    43, 45, 80, 117, 119
IMRO, see application software
IMS DB, see database software
IMS DC, see communications
    software
Induction, 44–8
information technology (IT), 6–8,
    34, 63, 100–3, 139–73,
    177–91, 193–213
    active vendors, 170–4

bespoke systems, 141–59
best-of-breed solutions, 181–4
industry response to MRO,
    139–75
legacy solutions, 178–81
leveraging, 63, 193–213
manage, 100–3
solutions and feasibility for
    MRO, 188–90, 199–207,
    207–10, 210–12
technologies, 186–8
infrastructure, 88, 102–3, 177
inspection, 51, 59, 80, 85, 97–8,
    108–9, 123, 129
installation, 106, 118, 129
Integrated Information
    Infrastructure – Reference
    Model (TOGAF III-TRM),
    102
interface, 103, 145, 146–7, 148–9,
    160, 178, 185–8, 203, 206
International Air Transport
    Association (IATA), 81, 82,
    90
International Business Machines
    (IBM), 33, 34, 35, 139–41,
    143, 144–51, 155, 160,
    163–5, 168, 170, 172
international logistics centre (ILC),
    121
internet, 119, 146–7, 163, 166,
    187, 203
interoperability, 55, 65, 72, 76
inventory, 29, 30, 83, 113
investment, 6, 16, 100, 141, 159,
    160, 168, 170, 174

Japan Airlines Ltd (JAL), 168
JAR 145, see standards

JCL, see programming language
Joint Engine Maintenance and
    Task Oriented Support System
    (JEMTOSS), 132, 133

Kennedy, President John F., 145
key performance indicators (KPIs),
    9
kits, 82
Kuwait Airways, 158

LAN-Chile, 172, 183
leasing, 18
liability, 27, 87–8
life cycle
    aircraft components
        maintenance, 71–5
    aircraft engine maintenance,
        64–71
    airframe maintenance, 53–64
    commercial aircraft, 36–53
    ground support equipment/fleet
        (GSE/F) maintenance, 75–8
log books, 44, 47, 134
logistics, 78–86, 9, 23, 24, 60, 68,
    74, 77–8, 85–6, 110, 112–3
LOT Polish Airlines, 158
low cost carriers (LCC), 12, 104
Lufthansa, 168, 179

mainframe, 6, 139, 141, 148, 149,
    154, 160, 162, 164, 173
Maintenance and Task Oriented
    Support System (MTOSS),
    132, 133
maintenance control centre
    (MCC), 59, 94
    Maintenance, Repair and
        Overhaul (MRO)
    and the IT industry, 115–37,
        140–1
    customers, 17–19, 30–1
    demand and capacity planning,
        19–20
    invoicing, 25–6
    IT landscape, 177–90
    IT solutions and feasibility,
        188–90, 199–207, 207–10,
        210–12
    market overview, 16–17
    orders, contracts and fulfilment,
        23–5
    service offerings, 20–3
    standards, 116–23
    technologies, 186–7
    warranty, 26–30
Maintenix, 168, 171–2
Manufacture
    and certification, 40–1
    aircraft parts, 105–6, 112
    original equipment
        manufacturer (OEM), 3, 5,
        11, 26–30, 42–3, 46–50, 60,
        78, 91, 104, 119–23, 127–8,
        134, 206
material specification data sheet
    (MSDS), 99
MAXI-MERLIN, see application
    software
MAXIMO, see application
    software
MAXIS, see application software
MEMIS, see application software
memory, 144, 150, 187
MERLIN, see application software
methodology, 34, 35
metrics, 57, 99
middleware, 139, 178

minimum equipment list (MEL), 10–11, 47
MRP–III, see standards
MRO, see Maintenance, Repair and Overhaul
MSG-2, see standards
MSG-3, see standards
MVS, see operating system
MXI Technologies of Canada, 167–8, 171–2, 183

National Fire Protection Association (NFPA), 95
Natural, See programming language
network, 102, 148
NFPA 409, see standards
non-destructive test (NDT), 65, 69
Northwest Airlines, 154, 155
NPL, see programming language

objectives, 1, 2, 7–8, 10, 98, 108
obsolescence, 52
operating system, 102, 144
  DOS, 142
  MVS, 142, 144, 146, 152, 154
  UNIX, 181, 186, 187, 203
  VM/VSE, 142, 154
  Windows, 186, 187
  Z/OS, 187
optimisation, 101–2, 121
ORACLE, 166–70, 174, 184, 185, 187, 199
original equipment manufacturer (OEM), 3, 5, 11, 26–30, 42–3, 46–50, 60, 78, 91, 104, 119–23, 127–8, 134, 206

Pan American Airlines (PanAm), 154
partnership, 145, 159, 165, 172
Peregrine, 170
performance, 28, 42, 57, 63, 80, 89, 97, 102, 105, 142, 144, 146, 152, 159, 161–2, 164, 180
personal computer (PC), 7, 160–1
phase-in, 44–8
phase-out, 52–3
PL/1, see programming language
planning, 5, 38, 43, 61, 81, 102, 140, 159, 164, 169, 178, 184, 188, 206, 210, 217
platform, 102, 103, 143, 155, 156, 159, 164–6, 173, 178
PMA, 105, 106, 110, 112, 114, 119
PricewaterhouseCoopers (PwC), 35, 168
procedure, 4, 6, 12, 38, 40, 52, 86, 90, 99, 106, 122, 132, 133
procurement, 24, 29, 49, 50, 60–1, 68, 74, 77–8, 80, 83, 113, 119, 122, 124, 133, 134
production, 55, 57, 61–2, 65, 67–9, 72, 76–8, 80, 84, 105, 120
program, 3–4, 7, 38, 96, 97, 109, 114, 118, 125, 146, 148, 180
programming language, 141, 142–3
  ASSEMBLER, 152
  COBOL, 141, 142, 143, 147, 152, 154, 165, 180
  FORTRAN, 143, 152

HTML, 43, 113, 118
JCL, 180
Natural, 156, 180
NPL, 143
PL/1, 165
SGML, 43, 118
XML, 43
protocol, 147–9, 152
provisioning, 48–50

QANTAS, 172, 174
quality, 111, 113

radar, 10, 27
radio-frequency identification
    (RFID), 6
RAMCO, 167, 168, 171, 172
Raptor, see application software
regulations, 13, 35, 85, 92, 95, 97,
    108–9, 119, 131
  compliance, 12–13
regulators, 11, 105, 119
reliability, 8, 69, 95–7, 102, 114,
    180
reliability-centred maintenance
    (RCM), 95, 96, 97
Republic Airlines, 153, 154
required inspection item (RII), 98,
    109
Resources, 61–2
responsibility, 22, 45, 47, 58,
    97–8, 100, 108, 156, 161,
    179
retirement, 52–3
revenue, 2, 12, 16, 26, 28, 29, 34,
    89, 111
Rockwell, 145
routing, 1, 18, 20, 22, 56
Royal Air Maroc, 158

SABRE, see application software
SAE JA1011, see standards
safety, 8–9, 13, 35, 46, 84, 91, 92,
    95–6, 98–9, 108–9, 114,
    128–9
SAP, 33–5, 46, 131, 166–7,
    168–70, 174, 187, 199
SCEPTRE, see application
    software
security, 9, 83, 103
server, 7, 44, 139, 160–4, 166,
    171, 173, 178
service bulletin (SB), 59
service oriented architecture
    (SOA), 190
servicing, 65, 72, 68, 70, 76
SGML, see programming language
signage, 99
significant item (MSI), 44, 126
Singapore International Airlines
    (SIA), 168
spacecraft, 93, 145
spares, 60–1
STA, 171
stakeholders, 31, 148
standards, 3, 12, 34, 35, 40, 43,
    46, 95, 106, 116, 117, 137
  AC 120-16D, 3, 11, 107, 109,
    115
  AC 120-17A, 95
  ATA 2100, 43
  ATA MSG-2, 21, 38, 132
  ATA MSG-3, 4–9, 10, 13, 21,
    37, 38, 40, 66, 76, 106,
    130–1
  EN 54, 95
  JAR 145, 3
  MRP-III, 106, 130–1
  NFPA 409, 95

iSPEC2200, 43
S1000D, 43
SAE JA1011, 97
statutes, 130
storage, 80–1, 93, 99, 124, 144, 145, 188
strategy, 1, 7, 9–11, 36, 44, 51, 58, 61, 80, 84, 96, 107, 109, 156, 167–8, 174
subtasks, 133
supplier furnished equipment (SFE), 120
supplier relationship (SRM), 45
SWISS, 171
systems integration, 115, 133–7, 156, 169, 177
Systems Network Architecture (SNA), 142, 148, 152

task card, 4, 60–1, 62–4, 67, 70, 73–4, 92, 132–3, 137, 162
telecommunications, 147–9
teletype, 148
telex, 83, 119
terminal, 142, 146, 147 148–51, 160
test, 65, 70, 72, 75, 76
The Open Group Architecture Framework (TOGAF) III-TRM, 102
training, 90, 91, 111, 133
transaction, 52, 81, 101, 103, 118–9, 121, 122, 141–2, 151

Transfer Control Protocol over Internet Protocol (TCP/IP), 147
TRAX, see application software
turnaround, 21–2, 81
type certificate data sheets database software (TCDS), 41

United Airlines, 95
UNIX, see operating system
USAir, 142, 153, 155

VisAer, 173, 174
VM, 154
VM/VSE, see operating system
volume, 7, 11, 14, 24, 43, 82, 100, 141
VSAM, see database software
VTAM, see communications software

warehousing, 78, 113, 131
warranty, 15, 25, 26–30, 71, 87–8
Windows, see operating system
work breakdown structures (WBS), 74, 78
workforce, 62, 91, 161
workstations, 142

XML, see programming language

Y2K (year 2000), 186

Z/OS, see operating system

Printed in the United States
By Bookmasters